# Excel Manual for

Moore's

# The Basic Practice of Statistics

*Second Edition*

Fred M. Hoppe

W. H. Freeman and Company
New York

Excel screen shots reprinted with permission from the Microsoft Corporation.



ISBN 0-7167-3611-X



Printed in the United States of America

Second printing, 2000

# Contents

# Preface

This book is a supplement to the second edition of *The Basic Practice of Statistics (BPS)* by David Moore. Its purpose is to present Excel as an approach to performing common statistical procedures.

Excel is a spreadsheet, a tool for organizing data contained in columns and rows. Operations on data mimic those described by mathematical functions, and the formulas required for data analysis can therefore be expressed as spreadsheet operations. Although originally developed as a business application to display numbers in a tabular format and to automatically recalculate values in response to changes in numbers in the table, current spreadsheets enjoy built-in functions and graphing capabilities that can be used for sophisticated statistical exploration and analyses.

At McMaster University, statistics is a prerequisite for many upper level courses. Students who complete a first course are expected to have practical experience in data analysis and to be able to apply their knowledge in laboratory or business courses involving real data.

Faced with a data set and a set of questions to be addressed, students are expected to know how to input the data into a software package, examine the data graphically for evidence that confirms or dispels their assumptions, draw justifiable conclusions, and finally export the results for presentation in report form.

In my large first-year course (800 - 1000 students per semester) I have begun to use Excel as the software of choice. There are a number of reasons for this. Excel is available at all campus workstations on a university wide site license. Many students have already learned to use Excel in the elementary or secondary grades and do not view it as a new piece of software to learn. Many have ready off-campus access to Excel, either at home, or as part of Microsoft Office, which they generally purchase for word processing and whose academic pricing meets their budgets. Excel has a familiar-looking windows interface which the students can understand and is capable of producing sophisticated graphs. Because Excel is part of an integrated word processing/graphics/database package, data can be easily input from other applications and exported into report form. For instance, I regularly download stock prices into Excel from the internet to plot the course of investments and my students enjoy this particular application as an incentive for learning statistics.

I believe that Excel can satisfy all of a student's needs in a course based on a book such as *BPS*. While it does not have the depth of specialized scientific software such as S-plus and Matlab, to which students are exposed in later years for advanced statistical and numerical computing, still every technique discussed in *BPS* can be readily developed within Excel. Besides, even software packages are not uniformly best suited for all purposes. For instance, although SAS can do logistic regression, the best tool for the task is S-plus.

Students who learn statistics using Excel will find the knowledge of Excel valuable in other courses (a point brought home by one of my third-year students who informed me that she used Excel in her physics course for plotting data because she "didn't feel like doing it by hand").

Moreover, in my own consulting I encounter researchers, who are self-taught in statistics, using Excel. Much of experimental research is exploratory in nature and Excel's excellent graphical capabilities and easy to use interface make it the statistics tool of choice for many in the life sciences, engineering, and business.

Finally, given that Excel is produced by Microsoft, one can predict without fear, a long useful and upgradeable life for Excel. All these considerations lead me to believe that Excel will grow in popularity in the teaching of statistics.

Some key features of this book are:

- Introductory chapter on Excel for those with no prior knowledge of Excel.
- Presentation follows *BPS* completely with fully worked and cross-referenced examples and exercises from *BPS*.
- Detailed exposition of the ChartWizard for graphical displays.
- Written for Excel 97/98 with an Appendix for Excel 5/95. Explanations are given in parallel for Excel 97/98 and Excel 5/95 when the interface differs.
- Completely integrated for use in Macintosh or Windows environment with equivalent keyboard and mouse commands presented as needed.
- Extensive use of Named Ranges to make formulas more transparent.
- More than 170 figures accompanied by step-by-step descriptions of all techniques.
- Figures from Excel 98 for Macintosh, Excel 97 for Windows, and Excel 5 for Macintosh showing differences and similarities in the user interface.
- Detailed use of simulation to explain randomness by simulating the Central Limit Theorem and the Law of Large Numbers.
- Construction of a normal table and graph of the normal density by a general method applicable to other densities.
- Templates provided for one-sample procedures, chi-squared tests, and nonparametric statistics.
- Development of side-by-side boxplots to supplement Excel.

The people at W.H. Freeman and Co. who were involved in this project were: Patrick Farace, Christopher Granville, Jodi Isman, and Christopher Spavins. My thanks to all. This book was completed while I was a member of the Fields Institute For Research in Mathematical Sciences whose excellent computing facilities I was able to use for this project.

I would like to thank my loving wife, Marla, for patiently putting up with my involvement in this book and for proofreading the entire manuscript thrice, and picking up many typos I missed. All remaining errors are my responsibility. My wonderful and dear children, Daniel and Tamara, helped as much as they could and enjoyed being part of the production process.

FRED M. HOPPE
DUNDAS, ONTARIO
JUNE 24, 1999.
E-mail: *hoppe@mcmaster.ca*
http://www.mathematics.net/

# Introduction

This book is a supplement to the second edition of *The Basic Practice of Statistics* by David S. Moore. Its purpose is to present Excel as an approach to performing common statistical procedures.

## I.1 What is Excel?

Microsoft Excel is a spreadsheet application whose capabilities include graphics and database applications. A spreadsheet is a tool for organizing data. Originally developed as a business application for displaying numbers in a table, numbers which were linked by formulas and updated whenever any part of the data in the spreadsheet changed, Excel now has built-in functions, tools, and graphical features that allow it to be used for sophisticated statistical analyses.

### Windows or Macintosh?

It doesn't matter which you use. This book is designed equally for Macintosh or Windows operating systems. The Macintosh and Windows versions of Excel function essentially the same way with a few slight differences in the file, print, and command shortcuts. These are due mainly to the lack of right mouse button for the Macintosh. However, the right button action can be duplicated with a keystroke and I have described both actions where they differ. Both Macintosh and Windows users will find this book equally useful.

Nearly all figures shown in this book have been taken using Excel 98 on a Macintosh running Mac OS 7.5.5. A few were from Excel 97 on a Toshiba notebook running Windows NT. The figures in the Appendix are from Excel 5 on a Macintosh. Students should feel familiar with the look of the Excel interface no matter what the platform.

### Which Version of Excel Should I Use?

Naming conventions are slightly confusing because of a plethora of patches, bug fixes, interim releases etc. In the Windows environment the main versions used are Excel 5.0, 5.0c, Excel 7.0, 7.0a (for Windows 95), Excel 8.0 (also called Excel

97), and the recent Excel 2000. For Macintosh, they are Excel 5.0, 5.0a and Excel 98.

The major change in the development occurred with Excel 5.0. The statistical tools in Excel 5.0 and Excel 7.0 for Windows, and Excel 5.0 for Macintosh function in virtually the same way. These are referred collectively in this book as Excel 5/95. Likewise Excel 97 for Windows and Excel 98 for Macintosh function similarly and these are referred as Excel 97/98.

Excel 97/98 removed some bugs in the Data Analysis ToolPak (but introduced others), improved the interface to the ChartWizard, and replaced the Function Wizard with the Formula Palette. Help was greatly expanded with the introduction of the animated Office Assistant. Minor cosmetic changes were made in the placement of components within some dialog boxes. Otherwise, few substantive changes occurred that might cause differences in execution or statistical capabilities between Excel 5/95 and Excel 97/98.

This book is based on Excel 97/98. However, given the large base of Excel 5/95 users I have tried to make this book equally accessible to them without having two separate editions. Surprisingly, this has not been too difficult. The main differences needing instructional care are the description of the ChartWizard (reduced to 4 steps in Excel 97/98 from 5 steps in Excel 5/95) and the implementation of formulas with the Formula Palette (Excel 97/98) instead of the Function Wizard (Excel 5/95). In addition, there are related differences in some pull-down options from the Menu Bar. On the other hand, the Data Analysis ToolPak is virtually the same in each version. Nearly all differences are in Chapters 1 and 2. Since this book uses a step-by-step approach with many figures to assist students visually in following written instructions, I have therefore included a detailed Appendix (with corresponding figures) which parallels for Excel 5/95 users the material in these chapters. In the remaining chapters, whenever there is a technique whose implementation varies between the two versions, I have given separate steps. Usually only a few lines of text suffice (four or less), underscoring the similarity in the versions with respect to the use of Excel for statistics. So while an instructor using this manual could make the necessary adjustments between versions, it is hoped that this dual approach will make this manual truly equally useful for Excel 5/95 and Excel 97/98.

## Do I Need Prior Familiarity With Excel?

The short answer is no. This book is completely self-contained. This introductory chapter contains a summary and description of Excel which should provide enough detail to enable a student to get started quickly in using Excel for statistical calculations. The subsequent chapters give step-by-step details on producing and embellishing graphs, using functions, and invoking the **Analysis Toolbox**. For users without prior exposure to Excel, this book may serve as a gentle exposure to spreadsheets and a starting point for further exploration of their features.

## I.2 The Excel Workbook

When you first open Excel, a new file is displayed on your screen. Figure I.1 shows an Excel 98 Macintosh opening workbook.

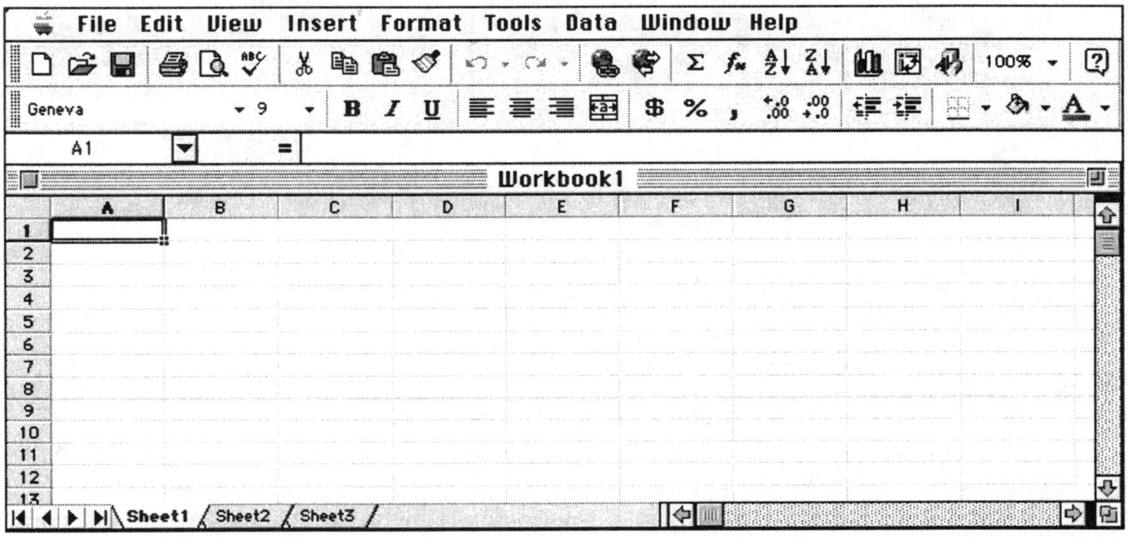

Figure I.1: Excel 98 Macintosh Workbook

A workbook consists of various sheets in which information is displayed, usually related information such as data, charts, or macros. Sheets may be named and their names will appear as tabs at the bottom of the workbook. Sheets may be selected by clicking on their tabs and may be moved within or between workbooks. To keep the presentation simple in this book we have chosen to use one sheet per workbook in each of our examples.

A sheet is an array of cells organized in rows and columns. The rows are numbered from 1 to 65,536 (up from 16,384 in Excel 5/95) while the columns are described alphabetically as follows:

A, B, C,..., X, Y, Z, AA, AB, AC,..., IV

for a total of 256 columns.

Each cell is identified by the column and row that intersect at its location. For instance, the selected cell in Figure I.1 has address (or cell reference) A1. When referring to cells in other sheets we need to also provide the sheet name. Thus Sheet2!D9 refers to cell D9 on Sheet 2.

Information is entered into each cell by selecting an address (use your mouse or arrow keys to navigate among the cells) and entering information, either directly in the cell or else in the **Formula Bar** text entry area. Three types of information can be entered: labels, values, and formulas. We will discuss entering information in detail later in this chapter.

Figure I.2: Excel 97 Windows

## I.3 Components of the Workbook

Look at Figure I.1. Then compare with Figure I.2 showing an Excel 97 (Windows) workbook and Figure I.3, an Excel 5 Macintosh workbook.

There are two main components in a workbook: the document window and the application window. Information is entered in the document window identified by the row and column labels. Above the document window are all the applications, functions, tools, formatting features that Excel provides. There are so many commands in a workbook that an efficient system is needed to access them. This is achieved either by menus with pull-down options invoked from the Menu Bar or from equivalent icons on the toobars.

The application window is thus the control center from which the user gives instruction to Excel to operate on the data in the document window below it. We examine the main components of the application window in detail.

### Menu Bar

The **Menu Bar** (Figure I.4) appears at the top of your screen. It provides access to all Excel commands: **File, Edit, View, Insert, Format, Tools, Data,**

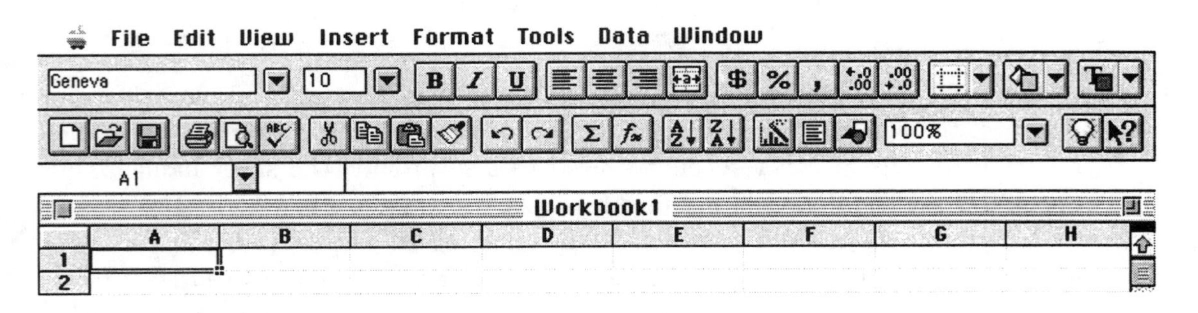


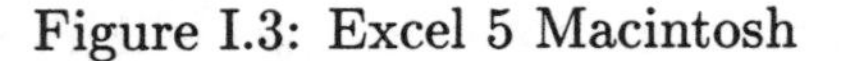
Figure I.3: Excel 5 Macintosh

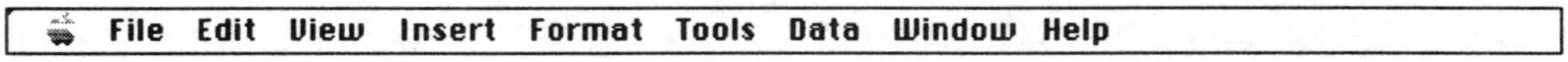

File Edit View Insert Format Tools Data Window Help

Figure I.4: Menu Bar

Table I.1: Menu Bar Pull-Down Options

| Menu Bar | Pull-Down Options |
|---|---|
| **File** | Open, close, save, print, exit |
| **Edit** | Copy, cut, paste, delete, find, etc. (basic editing) |
| **View** | Controls which components of workbook are displayed on screen, size, etc. |
| **Insert** | Insert rows, columns, sheets, charts, text, etc. into workbook |
| **Format** | Format cells, rows, columns |
| **Tools** | Access spelling macros, data analysis toolpak (will be used throughout to access the statistical features of Excel) |
| **Data** | Database functions such as sorting, filtering |
| **Window** | Organize and display open workbooks |
| **Help** | Online help (also available on the **Standard Toolbar**) |

**Window, Help**. Each word in the Menu Bar opens a pull-down menu of options familiar to users of any window-based application (there are also keyboard equivalents). As the name implies this is the main component of your control center and will be elaborated in various examples throughout this book. Table I.1 summarizes some of the options available.

### Toolbars

When Excel is opened two strips of icons appear below the Menu Bar: the **Standard Toolbar** and the **Formatting Toolbar**. Other toolbars are available by choosing **View – Toolbars** from the Menu Bar and making a selection from the choices available. Existing toolbars can be customized by adding or removing buttons and new ones can be created. For the purpose of this book you will not need to make such customizations.

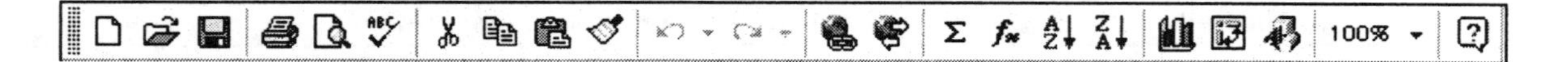

Figure I.5: Excel 98 Standard Toolbar

### Standard Toolbar

The (default) **Standard Toolbar** (Figure I.5) provides buttons to ease your access to basic workbook tasks. Included are buttons for the following tasks.

- Start a new workbook
- Open existing workbook
- Save open workbook
- Print, print preview
- Check spelling
- Cut selection and store in clipboard for posting elsewhere.
- Copy selected cells to clipboard, paste data from clipboard
- Copy format
- Undo last action, redo last action
- Insert hyperlink
- World-Wide Web interface
- Autosum function (may also be entered directly into cell or **Formula Bar**)
- **Paste Function** (step-by-step dialog boxes to enter a function connected to the **Formula Palette**)
- Sort descending, sort ascending order
- **ChartWizard** (covered in detail in Chapter 1).
- **PivotTable Wizard**
- Drawing toolbar
- Zoom factor for display
- **Office Assistant**

As you pass over a button with your mouse pointer a small text area appears next to the button describing its purpose.

## Formatting Toolbar

Figure I.6 shows the **Formatting Toolbar** by which you can change the appearance of text and data. Features offered include:

- Display and select font of selected cell
- Apply bold, italic, underline formatting
- Left, center, or right justify data)
- Merge and center

- Apply currency style, percent style, etc.
- Increase or decrease decimal places
- Indent
- Add borders to selected sides of cell
- Change background color of cell, change color of text in cells

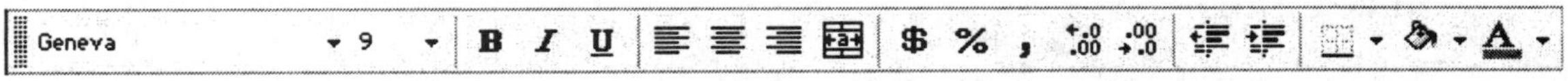

Figure I.6: Formatting Toolbar

## Formula Bar

The **Formula Bar** is located just above the document window (Figure I.7). There are six areas in the Formula Bar (from left to right):

Figure I.7: Formula Bar

- **Name box.** Displays reference to active cell or function.
- **Defined name pull down.** Lists defined names in workbook.
- **Cancel box.** Click on the red X to delete the contents of the active cell.
- **Enter box.** Click the green check mark to accept the formula bar entry.
- **Formula Palette.** (Excel 97/98) Constructs a function using dialog boxes to access Excel's built-in functions, or the function can be entered directly if you know the syntax. (Excel 5/95 has $f_x$ in place of the equal (=) sign to activate the **Function Wizard**).
- **Text/Formula Entry area.** Enter and display the contents of the active cell.

The **Cancel** box and **Enter** box buttons appear only when a cell is being edited. Once the data has been entered, they disappear.

## Title Bar

This is the name of your workbook. On a Mac the default is **Workbook1**, which appears just above the document window. With Windows the default name is **Book1**.

### Document (Sheet) Window

A sheet in a workbook contains 256 columns by 65,536 rows. Use the mouse pointer or arrow keys to move from cell to cell. The pointer may change appearance depending on what actions are permitted. It might be an arrow, a blinking vertical cursor (I-beam), or an outline plus sign, for instance.

### Sheet Tabs

A single workbook can have many sheets, the limit determined by the capacity of your computer and it is sometimes convenient to organize a workbook with multiple sheets, for instance sheets for data, for analyses, bar graphs, or for Visual Basic macros. Each sheet has a tab located at the bottom of the workbook (Figure I.8) and a sheet is activated when you click on its tab. Tab scrolling buttons allow you to navigate among the sheets. Clicking on a sheet tab on the bottom of the sheet activates it. Each workbook consists initially of 3 sheets labelled as Sheet1, Sheet2, Sheet3 (16 initially in Excel 5). Sheets can be added, deleted, moved, and renamed to achieve a logical organization of data and analyses. To rename, move, delete, or copy a sheet, **right-click (Windows)** on a sheet tab or click and hold down the Control key on a **Macintosh** – we will refer to this as **Control-click** – and a pop-up menu appears from which you can select. A new sheet can also be inserted from the Menu Bar by choosing **Insert – Worksheet**. Note that a new worksheet is added to the left of the current or selected sheet. Sheets can also be deleted, copied, or edited from the Menu Bar using **Edit – Delete Sheet** or **Edit – Move or Copy Sheet...** .

Another way to move a sheet is to grab it by clicking on it and holding the mouse button (left button for Windows). A small icon of a paper sheet will appear under the mouse pointer. As you move the mouse pointer, you will notice a small dark marker moving between the sheet tabs. This marker indicates where the sheet will be moved when you release the mouse button.

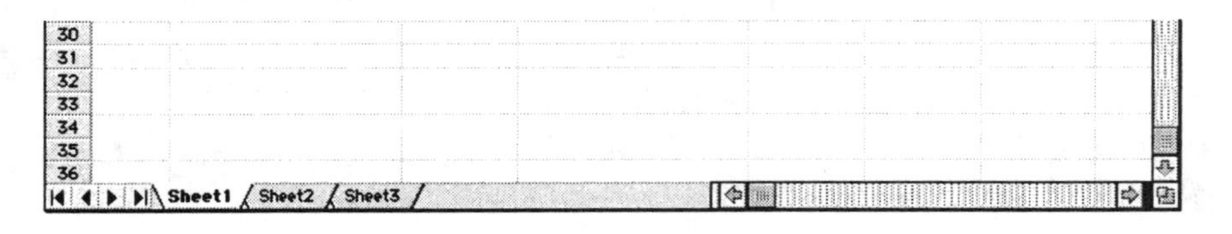


Figure I.8: Sheet Tabs

## I.4 Entering and Modifying Information

When a workbook is first opened cell A1 automatically becomes the active cell. Active cells are surrounded by a dark outline indicating they are ready to receive data. Use the mouse (or arrow keys on the keyboard) to activate a different cell. Then enter the data and either click on the Enter box or press the enter (return) key. Three types of information can be entered into a cell: labels, values, and formulas.

## Labels, Values, and Formulas

**Labels** are character strings such as words or phrases, typically used for headings or descriptions. They are not used in numerical calculations.

**Values** are numbers such as 1.3, \$1.75, $\pi$.

**Formulas** are mathematical expressions which use the values or formulas in other cells to create new values or formulas. All formulas begin with an equal (=) sign and are entered directly by hand in the cell, or in the text entry area of the **Formula Bar**, or by the **Function Wizard**. As an example of how a formula operates, if the formula

$$= A1 + A2 + A3 + A4$$

is entered in cell A5 and if the contents of cells A1, A2, A3, A4 are 11, 12, 19, −6 respectively, then cell A5 will show the value 36 because what is displayed in the cell is the result of the computation, not the formula. The formula in the cell may be viewed in the entry area in the **Formula Bar** if the cursor is placed over the cell.

It is the existence of formulas that makes a spreadsheet such a powerful tool. A formula such as

$$= \mathtt{SUM}(A1 : A4)$$

is the Excel equivalent of the mathematical expression

$$\sum_{i=1}^{4} A_i$$

and a complex mathematical expression can be rendered into an Excel workbook in a similar fashion.

## Editing Information

There are several ways to edit information. If the data has not yet been entered after typing then use the backspace or delete key, or click on the red X to empty the contents of a cell. After the data has been entered **activate** the cell by clicking on it. Then move the cursor to the text entry area of the Formula Bar where it turns into a vertical I-beam. Place the I-beam at the point you wish to edit and proceed to make changes.

## Cell References, Ranges, and Named Ranges

### Cell References

A **cell reference** such as A10 is a **relative** reference. When a formula containing the reference A10 is copied to another location, the cell address in the new location is changed to reflect the position of the new cell. For instance, if the formula presented earlier

$$= A1 + A2 + A3 + A4$$

which appears in cell A5 is copied to cell D9 then it will become

$$= D5 + D6 + D7 + D8$$

to reflect that the formula sums the values of the four cells just above its location.

This relative addressing feature makes it relatively easy to repeat a formula across a row or column of a sheet, such as adding consecutive rows, by entering the formula once in one cell then copying its contents to the other cells of interest.

If you need to retain the actual column or row label when copying a formula then precede the label with a dollar sign ($). This is called an **absolute** cell reference. For instance, $A2 keeps the reference to column A but the reference to row 2 is relative, A$2 leaves A as a relative reference, but fixes the row at 2, $A$2 gives the entire cell (row and column labels) an absolute reference. We will use mixed ($A2 or A$2) references in Chapter 9 with the chi-square distribution.

### Range

A group of cells forming a rectangular block is called a **range** and is denoted by something like A2:B4, which includes all the cells {A2 ,A3, A4, B2 B3, B4}.

### Named Ranges

**Named ranges** are names given to individual cells or ranges. The main advantage of named ranges is that they make formulas more meaningful and easy to remember. We will use named ranges repeatedly in this book. To illustrate, suppose we wish to refer to the range A2:A7 by the name "data." Enter the label "data" in cell A1. Then select the range A1:A7 by clicking on A1, holding the mouse button down, dragging to cell A7, and then releasing the mouse button. From the Menu Bar choose **Insert – Name – Create**, and check the box **Top Row** then click **OK** (Figure I.9).

Perhaps the quickest way to add a named range is to select the cell(s) to be named and then click on the **Name box**. This creates the name and associates it with the selected cells.

You can see which names are in your workbook by clicking on the defined name drop-down list arrow to the right of the **Name box**. Names are displayed alphabetically.

If you type a name from your workbook in the **Name** box and hit Enter then you will be transported to the first entry in the corresponding range, which will appear in the text/formula entry area of the Formula Bar ready for editing. Named ranges are both an aid in remembering and constructing formulas and also a convenient way to move around your workbook. They are used extensively in this book.

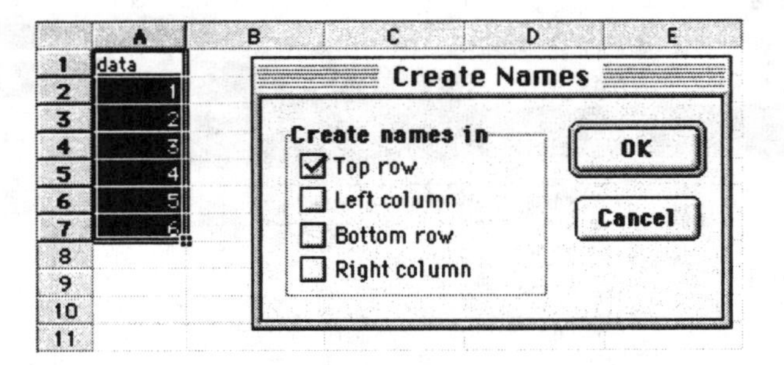


Figure I.9: Named Range

## Copying Information

To activate a block of cells place the cursor in the upper left cell of the block, click and drag to the lower right cell and release. Alternatively activate the upper left cell, then press the **Shift** key and click on the lower right cell. Non-contiguous cells or blocks can be selected by holding down the **Command** key **(Macintosh)** or the **Control** key **(Windows)** while selecting each successive cell or block.

To copy data from a cell or range, activate it, then from the Menu Bar choose **Edit – Copy**. Move the mouse cursor to the new location and from the Menu Bar choose **Edit – Paste**.

An alternative is to activate the range, then place the mouse cursor on the border of the selected range. It will appear as a pointer. Now press the **Control** key and move the cursor to the location for copying. Release the mouse button.

To move data to another location choose **Edit – Cut** from the Menu Bar and then **Edit – Paste** after you move the mouse cursor to the new location. Alternatively, move the mouse cursor to the border of the selected range. It turns into a pointer. "Grab" the border with the mouse pointer and move the cells to the desired location.

Use of the mouse for copying and cutting is a **drag and drop** operation familiar to users of Microsoft Word.

## Shortcut Menu

If you activate a range and then click on it with the **Control** key **(Macintosh)** pressed or with the **right mouse button (Windows)**, a **Shortcut Menu** will pop up next to the range allowing you access to some of the commands in the Menu Bar under **Edit**. This will provide options prior to copying or pasting.

### Paste Special

A useful command from the **Shortcut Menu** is the **Paste Special**. This provides a dialog box (Figure I.10) giving a number of options prior to pasting. The two options most commonly used in this book are **transpose** check box, which transposes rows and columns, and the **Paste Values** radio button, which is useful if you need to copy a range of values defined by formulas. A straight copy will alter

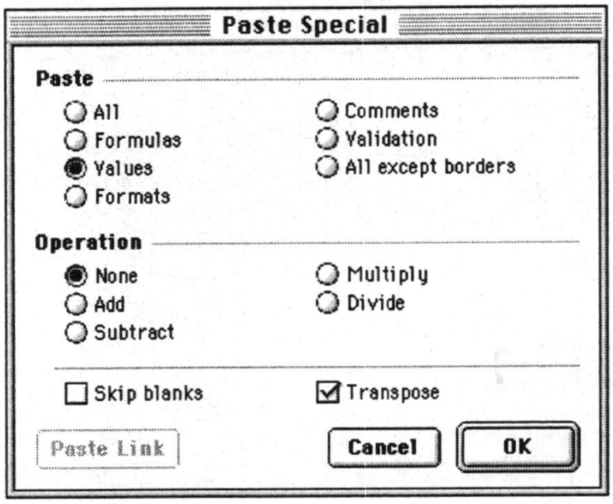

Figure I.10: Paste Special

the cell references in the formulas and could produce nonsense. **Paste Special** solves this problem by pasting the values, not the formulas.

### Filling

Suppose you need to fill cells A1:A30 with the value 1. Enter the value 1 in cell A1. Activate A1 and move the cursor to the lower right-hand corner (the **fill handle**) of A1. The cursor becomes a cross hair. Drag the fill handle and pull down to cell A30. This copies the value 1 into cells A2:A30.

Alternatively, you can select A1:A30 after entering the value 1 in cell A1. Then choose **Edit** – **Fill** – **Down** from the Menu Bar. Other options are available such as **Edit** – **Fill** – **Series** if this approach is taken.

## I.5 Opening Files

Often you will need to open text or data files. Other times, the data may be in a binary format produced by some other application.

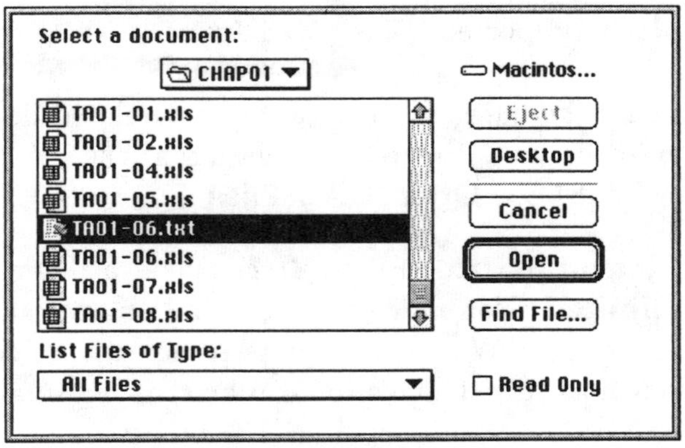

Figure I.11: Importing Files

### Binary

Excel can read and open a wide selection of binary files. To see which ones may be imported choose **File – Open** from the Menu Bar and make the appropriate selection from the drop-down list.

### Text (ASCII)

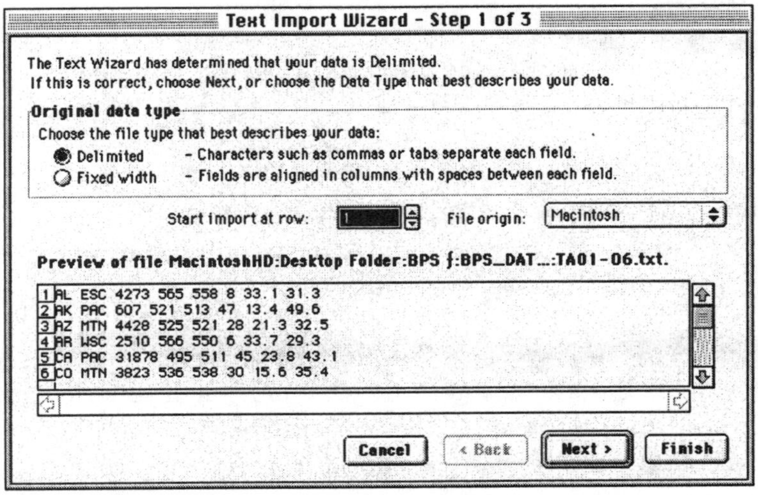

Figure I.12: Text Import Wizard – Step 1

For text (ASCII) files, Excel may start the **Text Import Wizard**, after you make your selection from the drop-down list, once you choose **File – Open** from the Menu Bar. This is a sequence of three dialog boxes allowing you to specify how the text should be imported. The **Text Import Wizard** helps you make intelligent choices.

We illustrate using the data set "Education and related data for the states" shown in Table 1.6 in the text. Figure I.11 shows that the file "TA01-06.txt" has been selected following **File** – **Open** – **All Files**.

**Step 1** The Text Import Wizard (Figure I.12) makes a determination that the data was **Delimited**. You may override this with the radio button. Sometimes the Text Import Wizard interprets incorrectly. The default starting position is shown as Row 1. This too can be changed if labels are needed.

**Step 2** The next screen depends on whether you chose Delimited or Fixed width in Step 1. If Delimited you pick the delimiter. If Fixed width you create line breaks.

**Step 3** The final step allows you to select how the imported data will be formatted. Usually the radio button **General**, the default, is appropriate.

In this case we accepted the defaults of the **Text Import Wizard** and the data was neatly imported. If the data file contains lines of explanatory text at the top of the file then the **Text Import Wizard** is very handy for formatting correctly upon import.

## I.6 Printing

This is generally the last step. In the **Page Setup** dialog box (Figure I.13) accessed from the Menu Bar using **File** – **Page Setup** there are four tabs:

- **Page.** Orientation, scaling, paper size, print quality.
- **Margins.** Top, bottom, left, right, preview window.
- **Header/Footer.** Information printed across top or bottom.
- **Sheet.** Print area, column/row title, print order.

The **Print Preview** button on the **Standard Toolbar** or **File** – **Print Preview** from the Menu Bar allows you to see what portion of your document is being printed and where it is positioned. Other options are available (Figure I.14) at the top of the print preview screen to assist you in previewing your output prior to printing.

Finally, when you are satisfied with your output press the **Print** button at the top of the preview screen. You can also print directly with the document window open using the **print** button on the **Standard Toolbar** or **File** – **Print** from the Menu Bar. This brings up a **Print** dialog box in which you select which pages to print, the number of copies, printer set up, as well as buttons for **Page Setup** and **Print Preview** just discussed.

Page Setup
Page | Margins | Header/Footer | Sheet
Orientation
Portrait
Landscape
Scaling
Adjust to: 100 % normal size
Fit to: 1 page(s) wide by 1 tall
First page number: Auto
Print quality: High
Print...
Print Preview
Options...
Cancel
OK

Figure I.13: Page Setup

Figure I.14:

## I.7 Whither?

This brief introductory chapter contains a bare-bones description of Excel, as much as you need to know to access the remainder of this book. In the following chapters not only will you make use of many of the topics and tips presented here, but you will learn about the **ChartWizard**, enter formulas, copy and paste cells, and so on. By using Excel for statistics you will also obtain a good practical background in spreadsheets.

One place to learn more about Excel is by using the animated **Office Assistant** as in Figure I.15. There is a gallery of different characters which you can install from the Excel installation CD in the Office:Actors folder. The one shown here is called **Max**. Max can be called up using the **Help** button (the ? at the right of the **Standard Toolbar**). Ask Max a question such as "What's new?" and see how he responds. If you **right-click (Windows)** or **Control-click (Macintosh)** Max, then more options become available.

There are numerous books available in libraries and in bookstores but these often contain too much information. Finally, there is a wealth of recent material

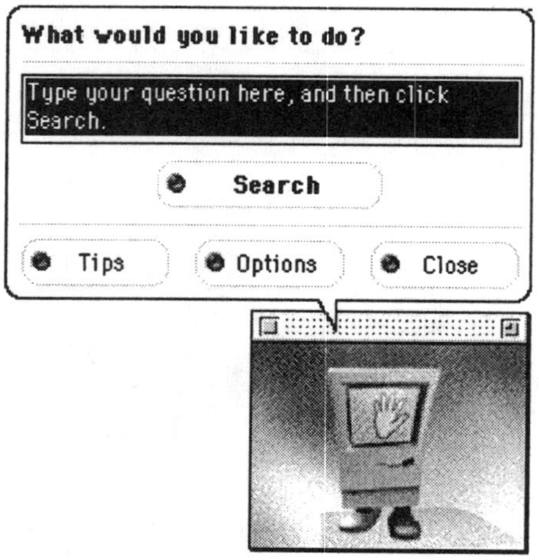

Figure I.15: Office Assistant – Max

available on the Internet. Use your favorite search engine or directory and you'll find pointers to macros (Visual Basic programs) and sample workbooks made available by other users for applications of Excel. Excel 97/98 is "internet ready" and can read html files and carry out "Web queries" to import data directly from the Internet.

If you find interesting Internet resources please let me know (*hoppe@mcmaster.ca*) and I will place a pointer to them on the Excel statistics web site which will be located at http://www.mathematics.net/.

# Chapter 1

# Examining Distributions

Data sets are measurements describing the variables associated with a group of individuals or objects. The distribution of a data set shows how often a particular measurement occurs and provides a complete description of the variation within the collection of measurements. From the distribution, useful global features may be obtained: graphical summaries, such as a histogram or scatterplot; or numerical summaries, such as a mean or standard deviation.

Such summaries are easily obtained in Excel, which provides more than 70 functions related to statistics and data analysis as well as tools in the **Analysis ToolPak** and the **ChartWizard**. The first tool we discuss is the **ChartWizard**. The figures are from Excel 98 for Macintosh. They differ only cosmetically from Excel 97 for Windows, as illustrated in Figure 1.1.

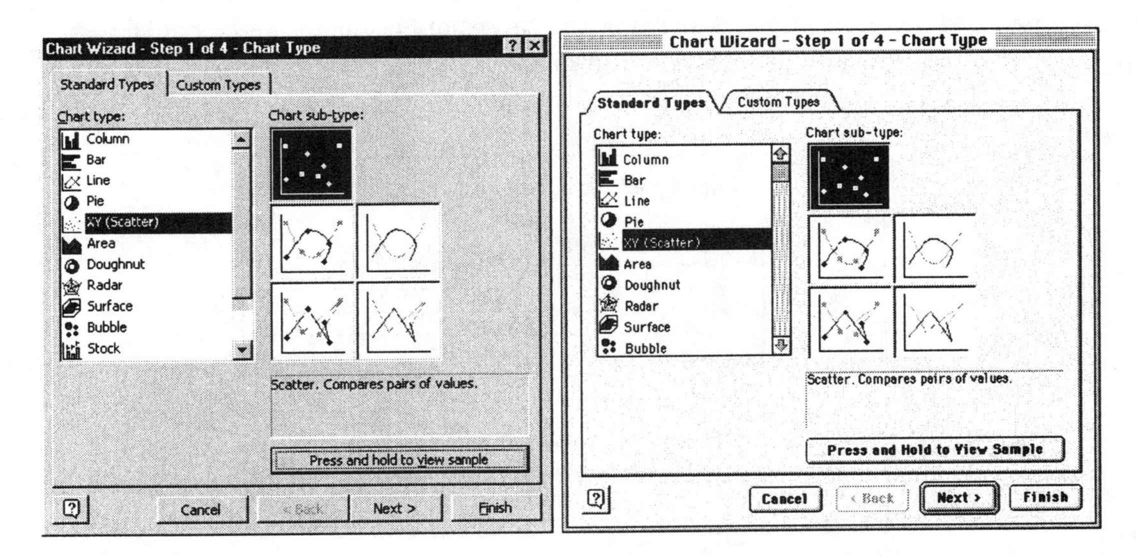


Figure 1.1: Comparison of Windows (left) and Macintosh (right) Screens

| | A | B |
|---|---|---|
| 1 | *Cause* | *Count* |
| 2 | auto | 43363 |
| 3 | falls | 10483 |
| 4 | drowning | 4350 |
| 5 | fires | 4235 |
| 6 | poison | 9072 |
| 7 | other | 18899 |

Figure 1.2: Bar Graph – Causes of Accidental Deaths

## 1.1 Displaying Distributions with Graphs

### The ChartWizard

The **ChartWizard** is a step-by-step approach to creating informative graphs. Its interface provides a sequence of four steps in **Excel 97 (Windows)** or **Excel 98 (Macintosh)** which guide the user through the creation of a customized graph (called a Chart by Excel). The user supplies details about the chart type, formatting, titles, legends, etc. in dialog boxes. The ChartWizard can be activated either from the button on the **Standard Toolbar** or by choosing **Insert – Chart** from the Menu Bar. The chart can be inserted in the current sheet or in a new sheet.

The following applies to Excel 97/98. **Users of Excel 5/95** should read Sections A.1 and A.2 in the Appendix instead.

> **Example 1.1.** (Exercise 1.4 page 8 in text.) Display the data in the workbook in Figure 1.2 showing the causes of accidental deaths in the United States in 1995 as a bar graph.

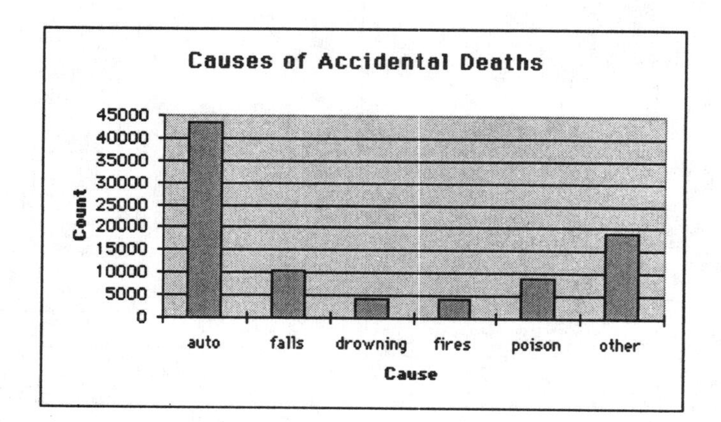


Figure 1.3: Excel Bar Chart

## Creating a Bar Chart

The steps for creating a bar graph are given below. For other types of graphical displays, make appropriate choices from the same sequence of dialog boxes.

Figure 1.3 is a bar graph produced by Excel which displays the same information as in Figure 1.2. The following steps describe how it is obtained. First enter the data and labels in cells A1:B7 and format the display as in Figure 1.2 for presentation purposes.

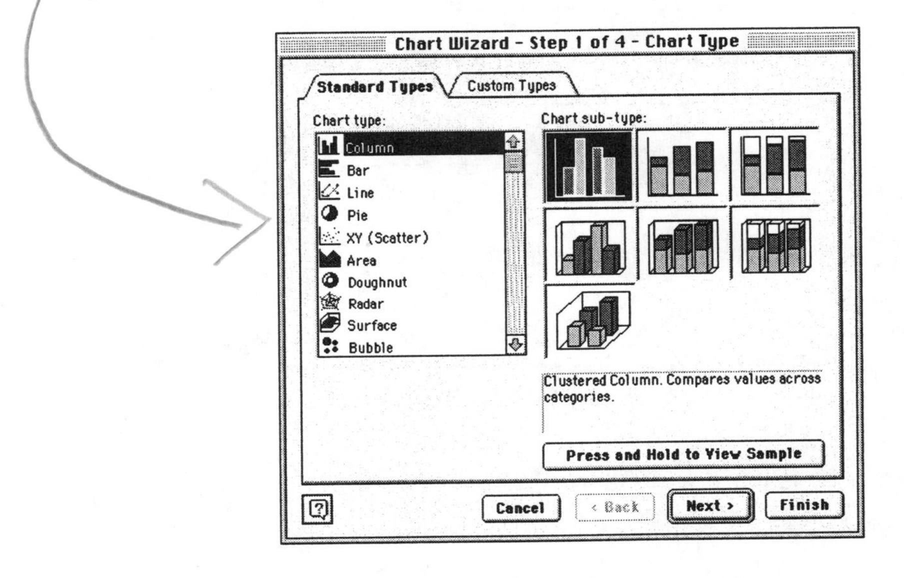


Figure 1.4: ChartWizard – Step 1

1. Step 1. Select cells A1:B7 and click on the **ChartWizard**. The ChartWizard (Figure 1.4) displays the types of graphs that are available. In the left field select **Column** for Chart type.

   In the right field select **Clustered Column** for Chart sub-type, which is the first choice in the top row. When you select a sub-type, an explanation of the chart appears in the box below all the choices and you can preview your chart's appearance using the "Press and hold to view your sample" button. Click Next.

2. Step 2. The next dialog box (Figure 1.5) with title **Chart Source Data** previews your chart and allows you to select the data range for your chart. Since you had already selected cells A1:B7 prior to invoking the ChartWizard, this block appears in the text area **Data range**. You can make any corrections. Had you not selected the data range then you would input the range now. Click Next.

3. Step 3. A dialog box **Chart Options** (Figure 1.6) appears with the default chart. Rarely is the default satisfactory and you will generally need to make cosmetic changes to its appearance.

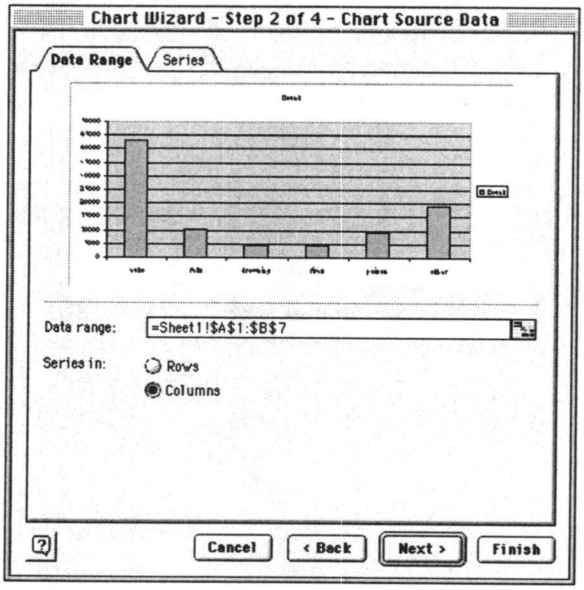

Figure 1.5: ChartWizard – Step 2

- Click the **Titles** tab. Enter "Causes of Accidental Deaths" for Chart title, "Cause" for Category (X) axis and "Count" for Value (Y) axis.
- Click the **Legend** tab. We don't require a legend since only one variable is plotted, so make sure the check box **Show Legend** is cleared.
- Additional tabs are available to customize other types of charts. They are not required here. Click Next.

4. **Step 4.** The final step lets you decide if you want the chart placed on the same worksheet as the data or in another worksheet. With each choice there is a field for entering the worksheet name. We will embed the chart on the same worksheet so we select the radio button **As object in:** (Figure 1.7). As our current workbook only has one sheet, Excel has used the default name Sheet1. We could also embed the chart on another sheet in the same workbook. Click the Finish button.

5. The chart appears with eight handles indicating that it is selected. The chart can be resized by selecting a handle and then dragging the handle to the desired size. The chart can also be moved. Click the interior of the chart and drag it to another location (holding the mouse button down). Then click outside the chart to deselect.

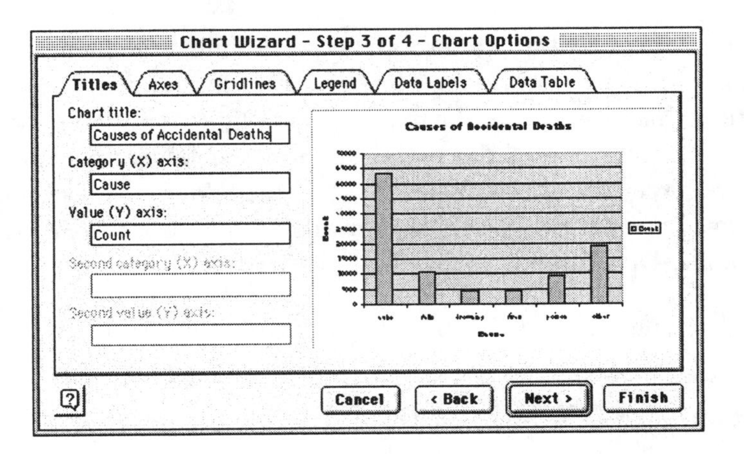


Figure 1.6: ChartWizard – Step 3

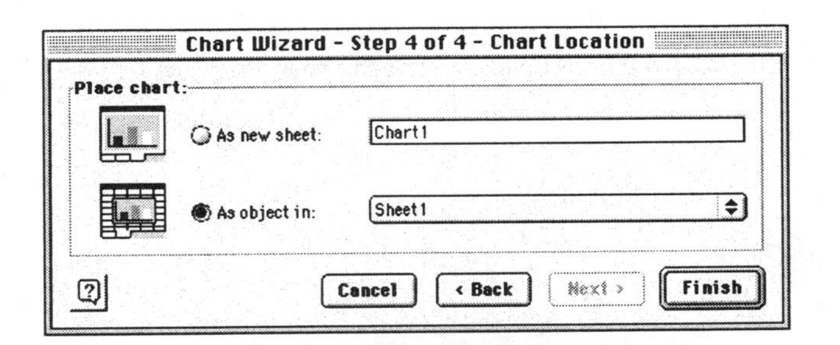


Figure 1.7: ChartWizard – Step 4

Figure 1.8: Chart Toolbar

You will also find the **Chart Toolbar** (Figure 1.8) embedded on the worksheet. This is used for embelishments of the chart. Use of the this toolbar is described in the next section on creating histograms. Note that the Chart Toolbar may also be called from the Menu Bar by **View – Toolbars – Chart**. Also, if you select the Chart by clicking once within its area, the Menu Bar will change. In place of the word **Data** there will now appear **Chart** from which a pull-down menu will provide the same tools as are displayed with icons on the Chart Toolbar.
**Note.** You might get an error message "Cannot add chart to shared workbook" even if you are not sharing your workbook. This is a bug in Excel 97/98 introduced when Shared Workbooks were implemented. It occurs under certain conditons if you try to create a chart using the data analysis tools.

You can still output your chart to a new workbook, and then, if desired, copy the chart to the existing workbook. This is one workaround.

Fortunately, this problem can be fixed by installing an updated file ProcDBRes to replace the existing one of the same name in the folder/directory "Microsoft Office 98:Office:Excel Add-Ins:Analysis Tools" (the folder for a default installation – your location may differ). This file is available for download at the Microsoft Software Library. Details may be found at the URL

http://support.microsoft.com/support/kb/articles/Q183/1/88.asp

## Pie Charts

It is easy as pie to produce a chart as in Figure 1.9 using Excel. Select Pie in place of Column in Step 1 of the ChartWizard and follow the remaining steps.

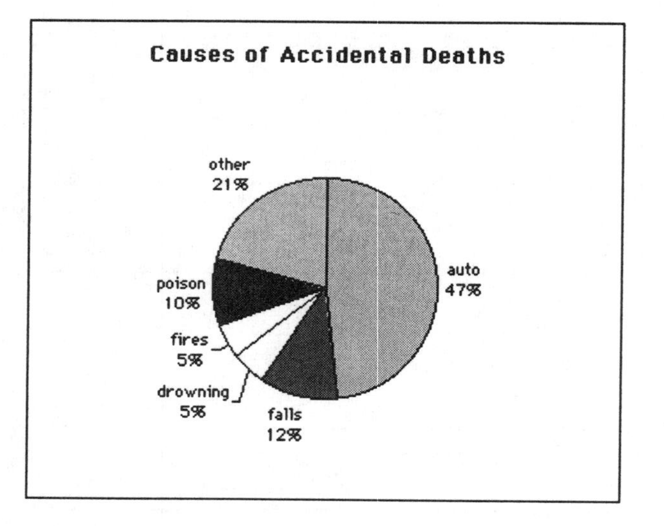


Figure 1.9: Pie Chart

Alternatively, since we have already created a bar graph, it is instructive to use the Chart Toolbar as an illustration of how easily modifications may be made. This interface is a vast improvement over the previous version of Excel.

1. Select the completed bar graph by clicking once within its border.
2. From the Menu Bar select **Chart – Chart Type. . . .** You will be presented with a box which is identical to Figure 1.4 but for the title which only contains the words **Chart Type** without mention of Step 1 of the ChartWizard. In the left field, referring to Figure 1.4, select **Pie** for Chart type and in the right field select **Pie** for Chart sub-type, which is the first choice in the top row.

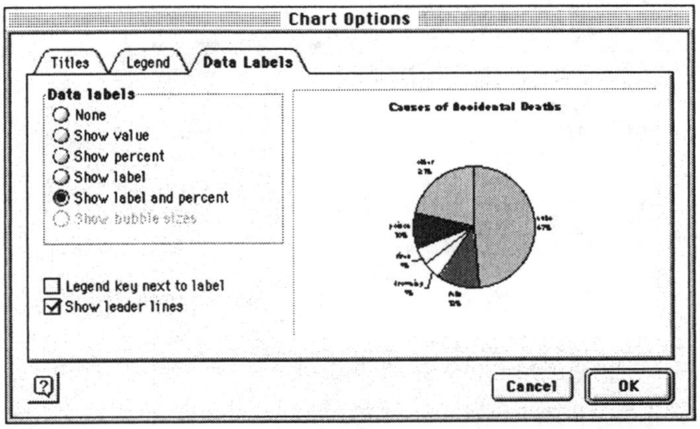

Figure 1.10: Chart Options

3. From the Menu Bar select **Chart – Chart Options. . . .** Figure 1.10 appears with three tabs Titles, Legend, and Data Labels. Click the tab **Data Labels** and then select the radio button **Show label and percent** and check the box **Show leader lines.** Click OK. A pie chart now replaces the bar graph.

## 1.2 Histograms

The ChartWizard is designed for use with data that is already grouped, for instance, categorical variables or quantitative variables which have been grouped into categories or intervals. For raw numerical data, Excel provides additional commands using the **Analysis ToolPak.**

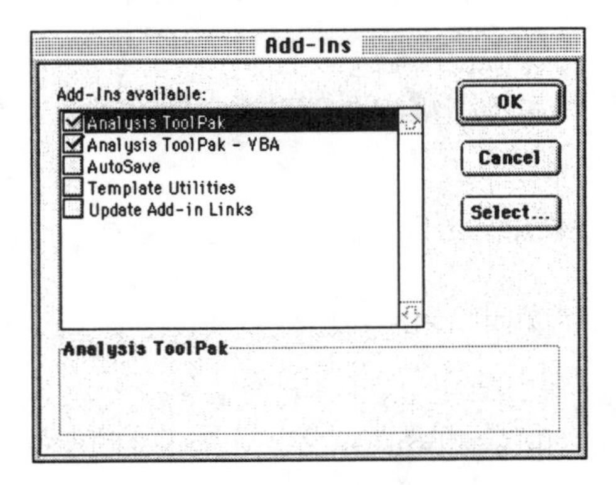


Figure 1.11: Data Analysis Add-In

To determine whether this toolpak is installed, choose **Tools – Add-Ins** from the Menu Bar. The **Add-Ins** dialog box (Figure 1.11) appears. Depending on

| | A | B | C | D | E | F | G | H | I | J |
|---|---|---|---|---|---|---|---|---|---|---|
| 1 | *Survival Times of Guinea Pigs* | | | | | | | | | |
| 2 | 43 | 45 | 53 | 56 | 56 | 57 | 58 | 66 | 67 | 73 |
| 3 | 74 | 79 | 80 | 80 | 81 | 81 | 81 | 82 | 83 | 83 |
| 4 | 84 | 88 | 89 | 91 | 91 | 92 | 92 | 97 | 99 | 99 |
| 5 | 100 | 100 | 101 | 102 | 102 | 102 | 103 | 104 | 107 | 108 |
| 6 | 109 | 113 | 114 | 118 | 121 | 123 | 126 | 128 | 137 | 138 |
| 7 | 139 | 144 | 145 | 147 | 156 | 162 | 174 | 178 | 179 | 184 |
| 8 | 191 | 198 | 211 | 214 | 243 | 249 | 329 | 380 | 403 | 511 |
| 9 | 522 | 688 | | | | | | | | |

Figure 1.12: Survival Times of Guinea Pigs

whether other Add-Ins have been loaded, your box might appear slightly different. If the Analysis ToolPak box is not checked, then select it and click OK. The **Analysis ToolPak** will now be an option in the pull-down menu when you choose **Tools – Data Analysis**. Note that you can also use the Select button to add customized add-ins to complement Excel.

## Histogram from Raw Data

**Example 1.2.** (Exercise 1.80 page 72 in text.) Make a histogram of survival times of 72 guinea pigs (Figure 1.12).

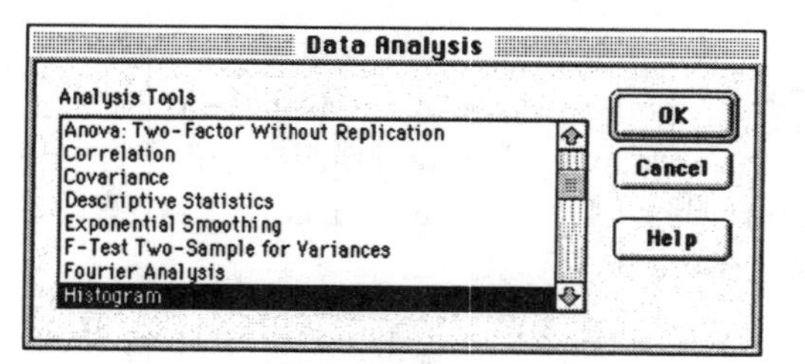


Figure 1.13: Data Analysis Tools

**Solution.** Excel requires a contiguous block of data for the histogram tool.

1. Enter the data in a block (cells A2:A73) with the label "Times" in cell A1.

2. From the Menu Bar choose **Tools – Data Analysis** and scroll to the choice Histogram (Figure 1.13). Click OK.

3. In the dialog box (Figure 1.14) type the reference for the range A1:A73 in the **Input range** area, which is the location on the workbook for the data. As with the Bar Chart you may instead click and drag from cell A1 to A73. The choice depends on whether your preference is for strokes (keyboard) or clicks (mouse). Leave the **Bin range** blank to allow Excel to select the bins,

check the **Labels** box because A1 has been included in the Input range, type C1 for **Output range**, to denote the upper left cell of the output range, and check the box **Chart output** to obtain a histogram on the same sheet of the workbook as the data. The option Pareto (sorted histogram) constructs a histogram with the vertical bars sorted from left to right in decreasing height. If Cumulative Percentage is checked, the output will include a column of cumulative percentages.

4. The output appears in Figure 1.15. The entries under Bin in C2:C10 are not the midpoints of the bin intervals, as you might expect. Rather they are the upper limits of the boundaries for each interval. The corresponding frequencies appear in cells D2:D10 with the histogram to the right. We shall shortly modify the histogram by changing the labels and allowing adjacent bars to touch. But first, we explain how to customize the selection of bins.

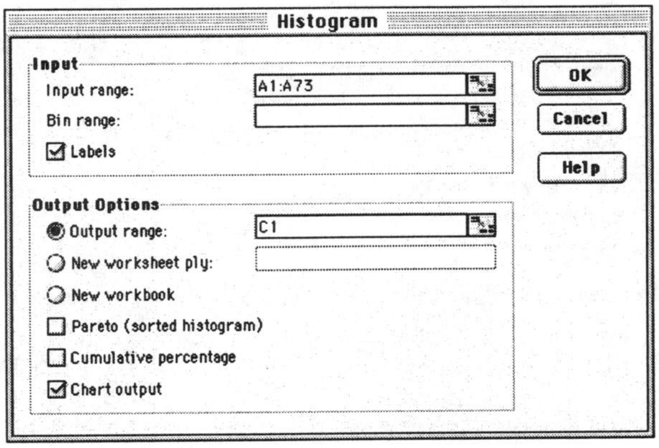

Histogram

Input
Input range: A1:A73
Bin range:
☑ Labels

Output Options
◉ Output range: C1
○ New worksheet ply:
○ New workbook
☐ Pareto (sorted histogram)
☐ Cumulative percentage
☑ Chart output

OK
Cancel
Help

Figure 1.14: Histogram Tool

| | C | D |
|---|---|---|
| 1 | *Bin* | *Frequency* |
| 2 | 43 | 1 |
| 3 | 123.625 | 45 |
| 4 | 204.25 | 16 |
| 5 | 284.875 | 4 |
| 6 | 365.5 | 1 |
| 7 | 446.125 | 2 |
| 8 | 526.75 | 2 |
| 9 | 607.375 | 0 |
| 10 | More | 1 |

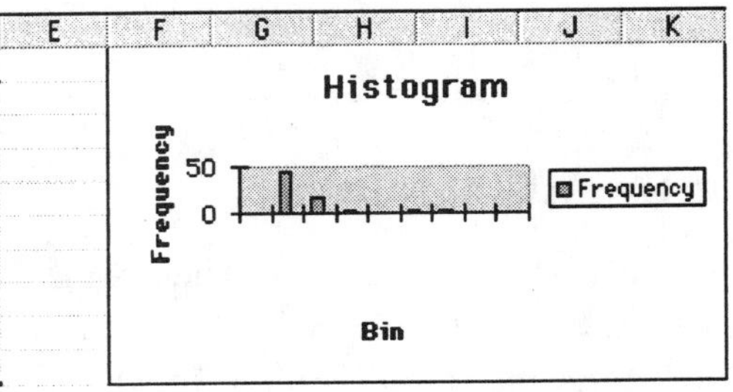

Histogram
Frequency
50
0
Bin
Frequency

Figure 1.15: Output Table and Default Histogram

## Changing the Bin Intervals

If the bin intervals are not specified then Excel creates them automatically, choosing the number of bins roughly equal to the square root of the number of observations beginning and ending at the minimum and maximum, respectively, of the data set. In the above example, we let Excel choose the default bins. Here we select our own bin intervals.

1. Type "New Bin" (or another appropriate label) in cell B1. Then enter the values 100, 200, 300, 400, 500, 600, and 700 in cells B2:B8. An easy way to accomplish this is to type 100 and 200 in cells B2 and B3 respectively, then select B2 and B3, click the fill handle in the lower right corner of B3, then drag the fill handle down to cell B8 and release the mouse button.

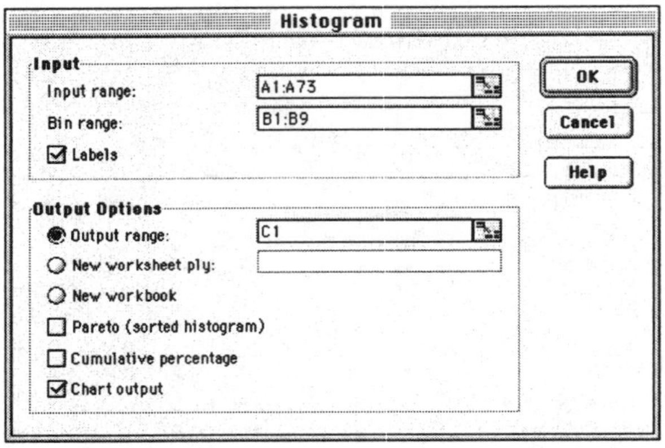

Figure 1.16: Histogram Tool – Specified Bin Intervals

2. Repeat the earlier procedure for creating a histogram, only this time type B1:B8 in the text area for **Bin range** (Figure 1.16). Before the output appears you will be prompted with a warning that you are overwriting existing data. Continue and the new output (Figure 1.17) will replace Figure 1.15 with your selected bin intervals.

## Enhancing the Histogram

While the default histogram captures the overall features of the data set, it is inadequate for presentation. Excel provides a set of tools for enhancing the histogram. These are too numerous to all be mentioned here, but a few will be discussed with reference to the example. The other options may be invoked analogously.

**Legend.** Remove the legend (which is not needed here) by clicking on it (the word "frequency" on the right in the chart in Figure 1.17), select **Chart – Chart**

| | C | D |
|---|---|---|
| 1 | *New Bin* | *Frequency* |
| 2 | 100 | 32 |
| 3 | 200 | 30 |
| 4 | 300 | 4 |
| 5 | 400 | 2 |
| 6 | 500 | 1 |
| 7 | 600 | 2 |
| 8 | 700 | 1 |
| 9 | 800 | 0 |
| 10 | More | 0 |

Figure 1.17: Output Table and New Bin Histogram

**Options...** from the Menu Bar, click the **Legend** tab, and clear the box **Show legend**. Alternatively you can choose **Edit – Clear – All** from the Menu Bar after selecting "frequency" on the chart.

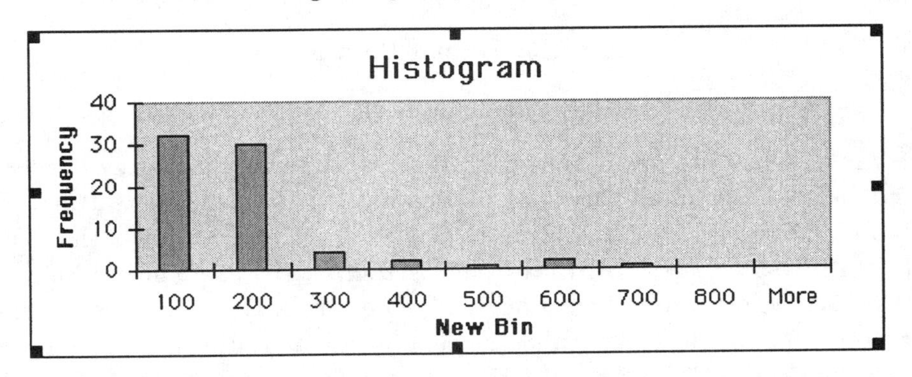


Figure 1.18: Resizing the Histogram

**Resize.** Both the histogram (called the **Plot Area**) and the bounding rectangle (called the **Chart Area**) which contains it can be resized and moved. Select the Chart Area by clicking once within its boundary resize using any of the eight the handles which appear or move it by dragging or cutting and pasting from the **Edit** on the Menu Bar to a new location. Likewise select the Plot Area by clicking once within its boundary and then resize or move as with the Chart Area. The X axis labels may appear placed horizontally, vertically or diagonally to accommodate the selected size. This can also be changed. After removing the Legend and resizing you will obtain something similar to Figure 1.18, which also shows the resize handles. Click **outside** the Chart Area to deselect.

**Bar Width.** Adjacent bars do not touch in the default, which looks more like a bar chart for categorical data. To adjust the bar width, double-click one of the bars to bring up the **Format Data Series** dialog box and select the Options tab. (Figure 1.19). Change the Gap width from 150% to 0%.

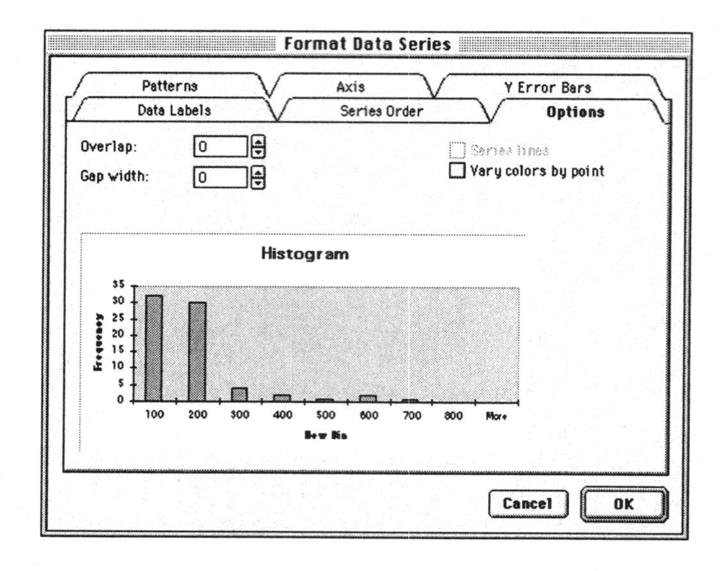


Figure 1.19: Format Data Series

**Chart Title.** Click on the title word Histogram. A rectangular grey border with handles will surround the word, indicating it is selected for editing. Begin typing "Survival Times (Days) of Guinea Pigs," hold down the **Alt** key (**Windows**) or the **Command** key (**Macintosh**), and press Enter. You may now type a second line of text in the **Formula Bar** entry area. Continue typing "in a Medical Experiment," then press Enter. If you want to move the title within the Chart Area use the handles. To change the font of your title, select the title, then from the Menu Bar choose **Format – Selected Chart Title. . . .** The dialog box has three tabs Patterns, Font, and Alignment. Select the **Font** tab and pick a font face, style and size.

**X axis Title.** Click on the word New Bin at the bottom of the chart and type "Survival Time (Days)." Change the font by selecting the X axis title, then from the Menu Bar choose **Format – Selected Axis Title. . .** and complete the dialog box as desired in the same fashion as for the Chart Title.

**Y axis Title.** Click on the word Frequency on the left side and then from the Menu Bar choose **Format – Selected Axis Title. . .** for any desired formatting.

**X axis Format.** Double-click the X axis and in the **Format Axis** dialog box (Figure 1.20) you can click on various tabs to change the appearance of the X axis. If you click on the **Alignment** tab you can change the orientation of the X axis labels.

**Y axis Format.** Double-click the Y axis and in the **Format Axis** dialog box you can click on various tabs to change the appearance of the Y axis. Click

Format Axis

Patterns | Scale | Font | Number | Alignment

Axis
Automatic
None
Custom
Style:
Color: Automatic
Weight:

Major tick mark type
None
Outside
Inside
Cross

Minor tick mark type
None
Outside
Inside
Cross

Tick mark labels
None
High
Low
Next to axis

Sample

Cancel
OK

Figure 1.20: Format X axis

on the **Scale** tab and change the **Maximum** to 40. Sometimes if you resize the chart you will need to change the **Major** or **Minor** units to achieve a pleasing result. Click OK.

**More Interval.** The More interval with 0 counts is unattractive. In the workbook in cell C10 (refer to Figure 1.17) change the label More to 900. The histogram is dynamically linked to the data in columns C and D and the label More on the X axis becomes 900. Of course, knowing that the counts are zero in the two bins labelled 800 and 900 we could redo the histogram ignoring those bins, if we wished to exclude them on the histogram.

**Plot Area Pattern.** The default histogram has a border around the Plot Area

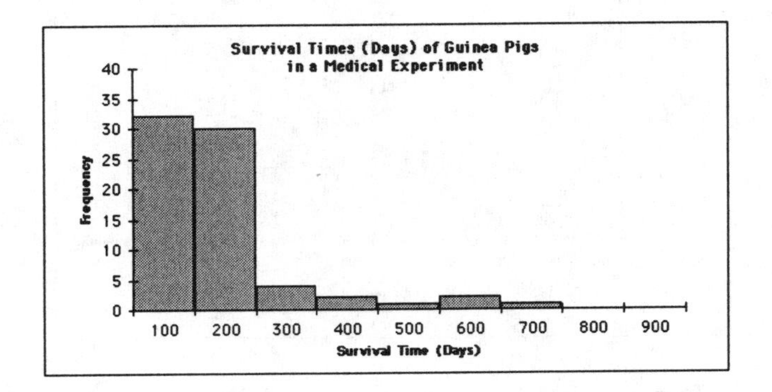


Figure 1.21: Final Histogram after Editing

and the Plot Area is shaded grey. Both defaults can be changed by double-clicking the **Plot Area** to bring up the **Format Plot Area** dialog box. Select the radio button **None** for Border and also select the radio button **None** for Area.

At the conclusion of the formatting, the histogram will look like Figure 1.21.

## Histogram from Grouped Data

The **Histogram** tool requires the raw data as input. When numerical data have already been grouped into a frequency table, it is the **ChartWizard** that is the appropriate tool. First use it to obtain a bar chart and then modify it exactly as you would enhance a histogram.

| | A | B | C |
|---|---|---|---|
| 1 | Class | Bin | Number of Students |
| 2 | 2.0 - 2.9 | 3 | 9 |
| 3 | 3.0 - 3.9 | 4 | 28 |
| 4 | 4.0 - 4.9 | 5 | 59 |
| 5 | 5.0 - 5.9 | 6 | 165 |
| 6 | 6.0 - 6.9 | 7 | 244 |
| 7 | 7.0 - 7.9 | 8 | 206 |
| 8 | 8.0 - 8.9 | 9 | 146 |
| 9 | 9.0 - 9.9 | 10 | 60 |
| 10 | 10.0 - 10.9 | 11 | 24 |
| 11 | 11.0 - 11.9 | 12 | 5 |
| 12 | 12.0 - 12.9 | 13 | 1 |
| 13 | Total | | 947 |

Figure 1.22: Vocabulary Scores – Grouped Data

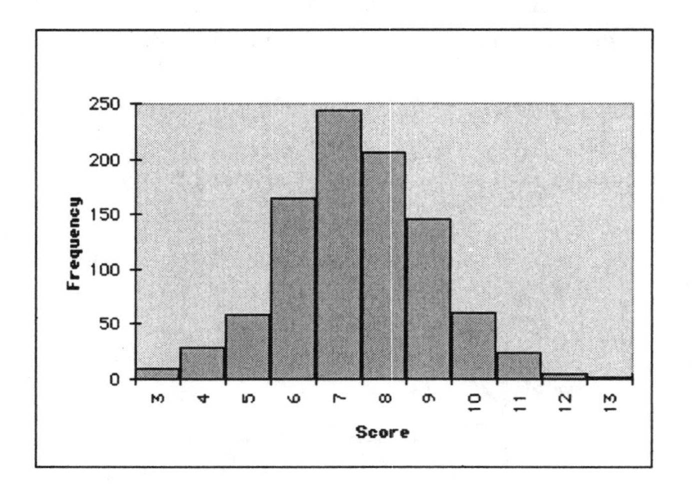


Figure 1.23: Histogram From Grouped Data

**Example 1.3.** (Figure 1.13 page 46 in text.) Figure 1.22 gives the frequencies of vocabulary scores of all 947 seventh graders in Gary, Indiana, on the vocabulary part of the Iowa Test of Basic Skills. Column A is the bin interval and column B is the label for the histogram. Construct a histogram using the **ChartWizard**. The final histogram is shown in Figure 1.23.

## 1.3 Describing Distributions with Numbers

| | A | B | C | D | E | F | G | H |
|---|---|---|---|---|---|---|---|---|
| 1 | Beef hot dogs | | | Meat hot dogs | | | Poultry hot dogs | |
| 2 | | | | | | | | |
| 3 | Calories | Sodium | | Calories | Sodium | | Calories | Sodium |
| 4 | 186 | 495 | | 173 | 458 | | 129 | 430 |
| 5 | 181 | 477 | | 191 | 506 | | 132 | 375 |
| 6 | 176 | 425 | | 182 | 473 | | 102 | 396 |
| 7 | 149 | 322 | | 190 | 545 | | 106 | 383 |
| 8 | 184 | 482 | | 172 | 496 | | 94 | 387 |
| 9 | 190 | 587 | | 147 | 360 | | 102 | 542 |
| 10 | 158 | 370 | | 146 | 387 | | 87 | 359 |
| 11 | 139 | 322 | | 139 | 386 | | 99 | 357 |
| 12 | 175 | 479 | | 175 | 507 | | 170 | 528 |
| 13 | 148 | 375 | | 136 | 393 | | 113 | 513 |
| 14 | 152 | 330 | | 179 | 405 | | 135 | 426 |
| 15 | 111 | 300 | | 153 | 372 | | 142 | 513 |
| 16 | 141 | 386 | | 107 | 144 | | 86 | 358 |
| 17 | 153 | 401 | | 195 | 511 | | 143 | 581 |
| 18 | 190 | 645 | | 135 | 405 | | 152 | 588 |
| 19 | 157 | 440 | | 140 | 428 | | 146 | 522 |
| 20 | 131 | 317 | | 138 | 339 | | 144 | 545 |
| 21 | 149 | 319 | | | | | | |
| 22 | 135 | 298 | | | | | | |
| 23 | 132 | 253 | | | | | | |

Figure 1.24: Hot Dog Data

The most direct way to obtain the common summary statistics is through the **Descriptive Statistics Tool**, which provides preformatted output very quickly. It is explained in this section. An alternative is the **Formula Palette** (replacing the **Function Wizard** of **Excel 5/95**) which provides greater flexibility of output and many more functions and formulas over its predecessor. We first describe the Descriptive Statistics Tool and then the Formula Palette.

## The Descriptive Statistics Tool

**Example 1.4.** (See Exercise 1.39 page 42 in text.) Figure 1.24 shows the calories and sodium levels measured in three types of hot dogs: beef, meat (mainly pork and beef), and poultry. Data is from *Consumer Reports*, June 1986, pp. 366-367. Describe the data using the Descriptive Statistics Tool.

**Solution.** For illustration purposes we only consider the beef calories data.

1. From the Menu Bar choose **Tools – Data Analysis** and double-click **Descriptive Statistics** (or, equivalently, select **Descriptive Statistics** and click OK) in the **Data Analysis Dialog** box. A dialog box **Descriptive Statistics** appears (Figure 1.25) which prompts for user input.

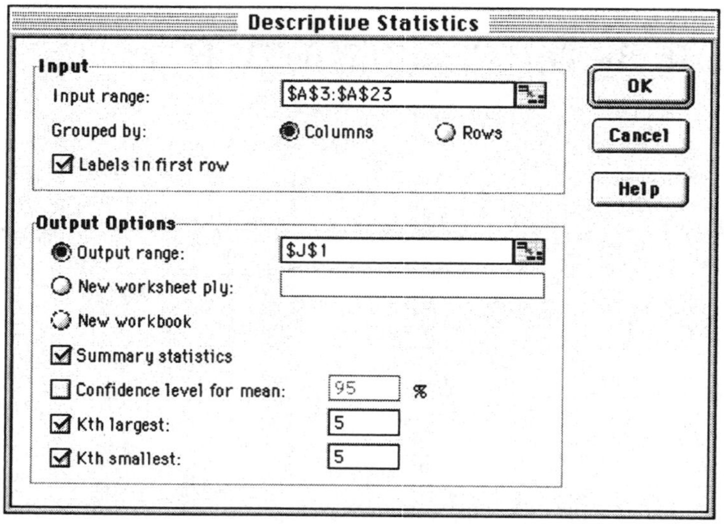

Figure 1.25: Descriptive Statistics Dialog Box

2. Complete the input as indicated in Figure 1.25. The **Input range:** is A3:A23 corresponding to the beef calories, including labels. (If you selected this range prior to invoking Descriptive Statistics, it will already be inserted by Excel.) Check the box **Labels in first row**. The **Confidence level for mean:** is not needed at this time (it gives the half-width). Check the **K-th largest:** or **K-th smallest:** boxes if needed. We have selected K = 5 for illustration.

3. The **Output Options** tell Excel where to place the output. Select cell J1. Finally check the box Summary Statistics and click OK. The output appears in Figure 1.26. We have formatted the output by reducing the number of decimal points using the **Decimal** button in the **Formatting Toolbar**. We can read off the summary statistics:

   mean = 156.85
   standard deviation = 5.06
   median = 152.50
   minimum = 111    maximum = 190
   5-th smallest = 139    5-th largest = 181

| | A | I | J | K |
|---|---|---|---|---|
| 1 | Beef hot dogs | | *Calories* | |
| 2 | | | | |
| 3 | Calories | | Mean | 156.85 |
| 4 | 186 | | Standard Error | 5.063 |
| 5 | 181 | | Median | 152.5 |
| 6 | 176 | | Mode | 149 |
| 7 | 149 | | Standard Devia | 22.642 |
| 8 | 184 | | Sample Varianc | 512.661 |
| 9 | 190 | | Kurtosis | -0.8131 |
| 10 | 158 | | Skewness | -0.0313 |
| 11 | 139 | | Range | 79 |
| 12 | 175 | | Minimum | 111 |
| 13 | 148 | | Maximum | 190 |
| 14 | 152 | | Sum | 3137 |
| 15 | 111 | | Count | 20 |
| 16 | 141 | | Largest(5) | 181 |
| 17 | 153 | | Smallest(5) | 139 |
| 18 | 190 | | | |
| 19 | 157 | | | |
| 20 | 131 | | | |
| 21 | 149 | | | |
| 22 | 135 | | | |
| 23 | 132 | | | |

Figure 1.26: Descriptive Statistics Output

## 1.4 The Formula Palette

In **Excel 97/98** the **Formula Palette** replaced **Function Wizard**. It assists in entering formulas and functions included in Excel, particularly complex ones. The functions can perform decision-making, action-taking, or value-returning operations. The Formula Palette simplifies this process by guiding you step by step.

It can be fired up in one of two ways. When you select a cell and press the **Paste Function** button $f_x$ next to the autosum button $\Sigma$ on the **Standard Toolbar** (or, equivalently, choose **Insert – Function...** from the Menu Bar) an equal sign (=) appears both in the cell and in the **Formula Bar**. The **Paste Function** dialog box (Figure 1.27) appears showing all available functions grouped by category on the left and the function name on the right. Both lists have scroll bars for choices not directly visible on the screen. At the bottom of the box, the selected function is shown with the arguments it takes and a brief description. (In previous versions of Excel a similar dialog box called **Function Wizard – Step 1 of 2** appeared.) When you click OK in the **Paste Function** box, the **Formula Palette** box appears below the **Formula Bar** requesting parameters and an input range for the function you selected. In addition, the Formula Bar is now activated showing the Formula Palette's drop-down list control with the 10 most recently used functions, and an equal (=) sign appears in the Formula Bar showing the selected function partially constructed and awaiting completion of its arguments. You may enter these either directly into the Formula Bar or in the Formula Palette box.

However, the **Formula Palette** is usually invoked more in a second, more

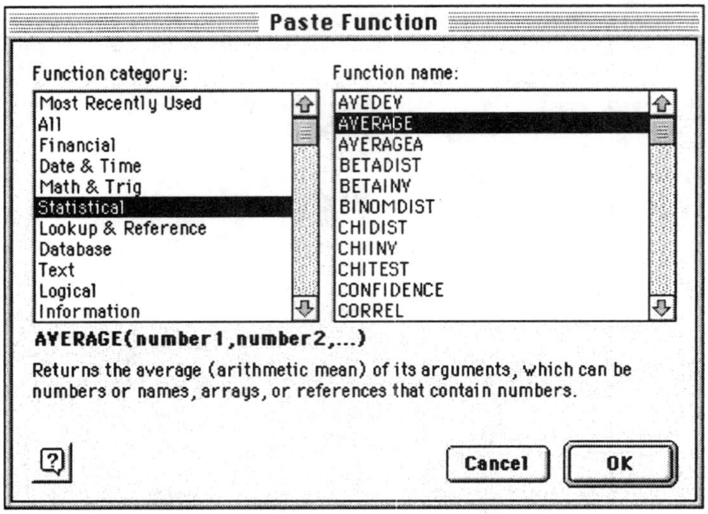

Figure 1.27: Paste Function

Figure 1.28: Formula Palette – Default

direct way. Select a cell and press (=) on the **Formula Bar** to open the **Formula Palette** dialog box (Figure 1.28). On the far left side of the **Formula Toolbar** is a button with the most recently used function, in this case `AVERAGE`. If this is the function you need then click on the word `AVERAGE` and the **Formula Palette** dialog box will expand (Figure 1.29) requesting the required parameter or the data range for the function (which can be typed directly or *referenced* by using the mouse to point to the data by clicking and dragging over cells in the data range). As you input this information, Excel will correspondingly build the function both in the **Formula Bar** and in the cell you had selected in the workbook. When you have completed entering the requested input, click OK to complete the function. If you want some other function than the default, click the small arrow to the right of the function name. Select from the drop-down list of your 10 most recently used functions or select **More functions....** If you select the latter then the **Paste Function** dialog box encountered above appears. An OK (Checkmark symbol) and Cancel (an X) button appear to the right of this arrow. Click the Checkmark and the formula is entered into the active cell. Click the Cancel to discard the formula without making changes.

**Recommendation.** The **Paste Function** button on the Standard Toolbar duplicates the actions of the **Formula Palette**. Since Excel formulas start with

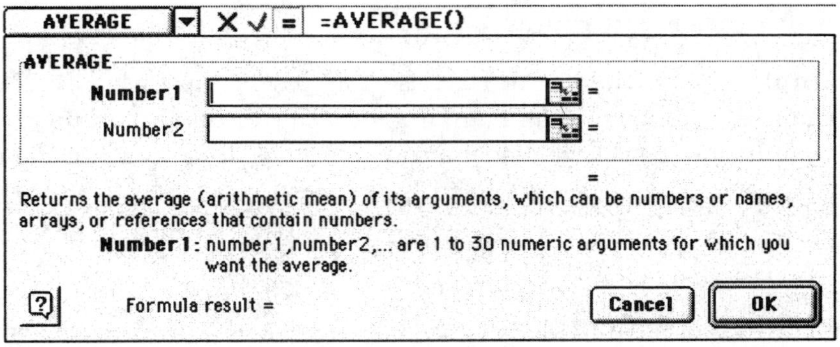

AVERAGE =AVERAGE()

AVERAGE

**Number1** =

Number2 =

=

Returns the average (arithmetic mean) of its arguments, which can be numbers or names, arrays, or references that contain numbers.

**Number1:** number1,number2,... are 1 to 30 numeric arguments for which you want the average.

Formula result =

Cancel OK

Figure 1.29: Formula Palette – Expanded

an (=) sign we recommend that you begin your formulas by pressing the (=) on the Formula Toolbar instead of using Paste Function. This activates the **Formula Palette** and you can either type the formula by hand into the Formula Bar or order up a function from the **Paste Function** box, if required. Besides, experienced users of Excel often **customize** the **Standard Toolbar** and replace the Paste Function button with some other one.

We illustrate use of the **Formula Palette** by deriving the five-number summary of a data set.

| | A | B | C | D |
|---|---|---|---|---|
| 1 | Beef Hot Dogs | | *Five Number Summary* | |
| 2 | | | | |
| 3 | Calories | | | |
| 4 | 186 | Min | 111 | =MIN(A4:A23) |
| 5 | 181 | Q1 | 140.5 | =QUARTILE(A4:A23,1) |
| 6 | 176 | Med | 152.5 | =MEDIAN(A4:A23) |
| 7 | 149 | Q3 | 177.25 | =QUARTILE(A4:A23,3) |
| 8 | 184 | Max | 190 | =MAX(A4:A23) |
| 9 | 190 | | | |
| 10 | 158 | | | |
| 11 | 139 | | | |
| 12 | 175 | | | |
| 13 | 148 | | | |
| 14 | 152 | | | |
| 15 | 111 | | | |
| 16 | 141 | | | |
| 17 | 153 | | | |
| 18 | 190 | | | |
| 19 | 157 | | | |
| 20 | 131 | | | |
| 21 | 149 | | | |
| 22 | 135 | | | |
| 23 | 132 | | | |

Figure 1.30: Five-Number Summary

## The Five-Number Summary

**Example 1.4.** (Data from Exercise 1.39 page 42 in text.) Find the five-number summary {minimum, first quartile, median, third quartile, maximum} for the calorie distribution of the hot dogs shown in Figure 1.30.

**Solution.**

1. Enter the labels "Min," "Q1," "Med," "Q3," and "Max" as shown in cells B4:B8 (Figure 1.30).

2. Click the equal (=) sign on the **Formula Bar** to start the **Formula Palette** and use the drop-down list to select **More functions....** In the ensuing **Paste Function** dialog box select **Statistical** from the left and scroll down and select `QUARTILE` on the right. Click OK.

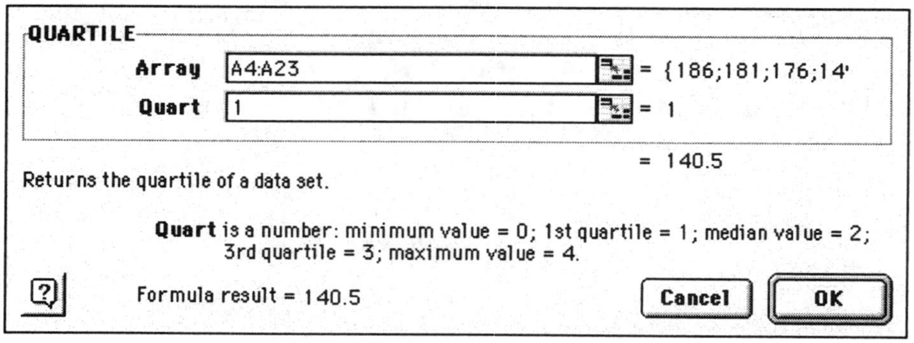

QUARTILE
Array A4:A23 = {186;181;176;14
Quart 1 = 1
= 140.5
Returns the quartile of a data set.
**Quart** is a number: minimum value = 0; 1st quartile = 1; median value = 2; 3rd quartile = 3; maximum value = 4.
Formula result = 140.5
Cancel OK

Figure 1.31: Quartile Formula

3. The **Formula Palette** dialog box appears. Move it out of the way and enter the data **Array** by selecting cells A4:A23 with your mouse (or more mundanely by typing A4:A23 into the dialog box). Click in the text area for **Quart** and type "1" to indicate the first quartile. The completed formula appears in the **Formula Toolbar** and value of the formula 140.5 shows in the dialog box (Figure 1.31). Click OK and the value 140.5 is printed in C5.

4. Continue in this fashion using the Formula Palette to complete the five number summary. Of course you can still enter the formulas by hand in the Formula Bar once you are familiar with them. Cells C4:C8 present the syntax while the values are in B4:B8.

The five-number summary is {111, 140.5, 152.5, 177.25, 190}. Note that Excel uses a slightly different definition of quartiles for a finite data set than the text.

## 1.5 The Normal Distribution

Areas under a normal curve can be found using the `NORMDIST` function. The syntax is: = `NORMDIST`($x, \mu, \sigma$, cumulative) where $\mu$ is the mean and $\sigma$ is the standard deviation. The parameter cumulative indicates whether the density (set cumulative = "false" or "0") or whether the cumulative distribution (set cumulative = "true" or "1") is wanted. The formula = `NORMDIST`($x, \mu, \sigma, 1$) returns $F(x)$ which is the area to the left of $x$ under an $N(\mu, \sigma)$ density and can be used to produce a table of normal areas as found in many statistics texts. Another formula = `NORMINV`($p, \mu, \sigma$) returns the inverse $F^{-1}(x)$ of the cumulative that is a value $x$ such that the area to the left of $x$ is the specified $p$. For $N(0,1)$ use `NORMSDIST` and `NORMSINV` instead.

## Normal Distribution Calculations

**Example 1.5.** (Examples 1.16 and 1.17 page 59 in text.) The level of cholesterol in the blood is important because high cholesterol levels increase the risk of heart disease. The distribution of blood cholesterol levels in a large population of people of the same age and sex is roughly normal. For 14-year-old boys the mean $\mu = 170$ milligrams of cholesterol per deciliter of blood (mg/dl) and the standard deviation is $\sigma = 30$ mg/dl. Levels above 240 mg/dl may require medical attention.
(a) What percent of 14-year-old boys have more than 240 mg/dl of cholesterol?
(b) What percent of 14-year-old boys have blood cholesterol between 170 and 240 mg/dl?

1. Click on a cell (activate it) where you want to locate the answer, cell A1.

2. The syntax is

$$\Phi(x) = \texttt{NORMDIST}(\mu, \sigma, x, 1)$$

so enter the formula $= 1 -$ `NORMDIST`(240, 170, 30, 1) since the upper tail area is wanted. The answer 0.00982 appears in cell A1. (Users of **Excel 5/95** may also use the **Function Wizard** while users of **Excel 97/98** may use the **Formula Palette** instead to enter the formula.)

**Example 1.6.** (Example 1.18 page 61 in text.) Scores on the SAT verbal test in recent years follow approximately the $N(505, 110)$ distribution. How high must a student score in order to place in the top 10% of all students taking the SAT?

**Solution.**

1. Activate cell B1 for entry of the function.

2. The syntax is

$$\Phi^{-1}(x) = \texttt{NORMINV}(x, \mu, \sigma)$$

for the inverse of the cumulative normal distribution. Enter the formula $=$ NORMINV(0.90, 505, 110) and read off 645.971, the 90th percentile of the SAT scores.

| | A | B | C | D | E |
|---|---|---|---|---|---|
| 1 | *Calculating Normal Areas and Inverse Probabilities* | | | | |
| 2 | *Parameters* | | | | |
| 3 | mean | | | | |
| 4 | sigma | | | | |
| 5 | | | | | |
| 6 | *Cumulative* | | | | |
| 7 | x | | | | |
| 8 | Area to left of x | =NORMDIST(B7, B3, B4, 1) | | | |
| 9 | Area of right of x | | | | |
| 10 | Standard Score Z | | | | |
| 11 | | | | | |
| 12 | *Interval* | | | | |
| 13 | a | | | | |
| 14 | b | | | | |
| 15 | Area between a and b | =NORMDIST(B14, B3, B4, 1)-NORMDIST(B13, B3, B4, 1) | | | |
| 16 | | | | | |
| 17 | *Inverse* | | | | |
| 18 | area A to left | | | | |
| 19 | percentile | =NORMINV(B18,B3,B4) | | | |

Figure 1.32: Calculating Normal Areas and Percentiles

Problems involving $\Phi(x)$ and $\Phi^{-1}(x)$ occur repeatedly, and it may be convenient to use a template. Figure 1.32 shows a workbook with the required formulas.

## Graphing the Normal Curve

By combining the **ChartWizard** and the NORMDIST function, we can create a graph of any normal curve. *In fact, the procedure described below can be used to plot the graph of any function which Excel can evaluate.*

### Constructing a Graph of the Standard Normal Curve

1. Enter the labels $z$ and $f(z)$ in cells A2, B2. (See Figure 1.33.)

2. Enter $-3.5$ and $-3.4$ in cells A3 and A4 respectively. We are going to create a column of $z$ values at which the standard normal density will be calculated. Select A2:A3, check the fill handle in the lower right corner of A3, and drag to cell A73 to fill the column with decreasing values of $z$ decremented by 0.1. Format the values with two decimal places.

3. Select cell B3 and enter = NORMDIST(A3, 0, 1, 0) in the **Formula Bar**. Cell B3 now contains the value 0.00087268, the standard normal density evaluated at $z = -3.50$.

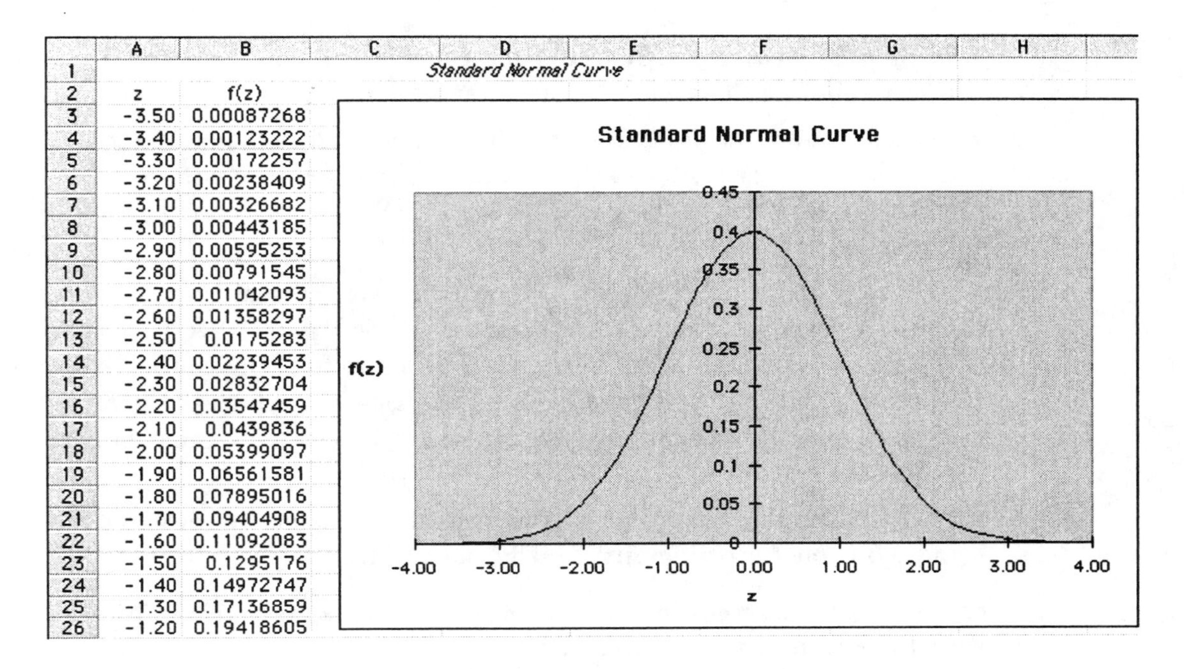

| | A | B |
|---|---|---|
| 1 | | |
| 2 | z | f(z) |
| 3 | -3.50 | 0.00087268 |
| 4 | -3.40 | 0.00123222 |
| 5 | -3.30 | 0.00172257 |
| 6 | -3.20 | 0.00238409 |
| 7 | -3.10 | 0.00326682 |
| 8 | -3.00 | 0.00443185 |
| 9 | -2.90 | 0.00595253 |
| 10 | -2.80 | 0.00791545 |
| 11 | -2.70 | 0.01042093 |
| 12 | -2.60 | 0.01358297 |
| 13 | -2.50 | 0.0175283 |
| 14 | -2.40 | 0.02239453 |
| 15 | -2.30 | 0.02832704 |
| 16 | -2.20 | 0.03547459 |
| 17 | -2.10 | 0.0439836 |
| 18 | -2.00 | 0.05399097 |
| 19 | -1.90 | 0.06561581 |
| 20 | -1.80 | 0.07895016 |
| 21 | -1.70 | 0.09404908 |
| 22 | -1.60 | 0.11092083 |
| 23 | -1.50 | 0.1295176 |
| 24 | -1.40 | 0.14972747 |
| 25 | -1.30 | 0.17136859 |
| 26 | -1.20 | 0.19418605 |



Figure 1.33: Graphing a Normal Density

4. Select cell B3, click the fill handle, and drag down to B73. This copies the formula you just entered in B3 into cells B4:B73 relative to the corresponding cell references in column A. Column B is filled with values $f(z)$ of the standard normal density corresponding to each value of $z$ in column A.

5. **Users of Excel 5/95.** Select cells A2:B73, click the **ChartWizard** button and then click in cell C2 and drag to I26 to locate the graph. A dialog box **ChartWizard – Step 1 of 5** appears.

   - In Step 1 you are given an opportunity to correct or confirm your range.
   - In Step 2 select **XY (Scatter)** chart.
   - In Step 3 select format **6**.
   - In Step 4 select the radio button **Columns** for Data Series, enter "1" for Column for Category (X) Axis Labels, and enter "1" for Row for Legend Text.
   - In Step 5 select the radio button **No** for Add a Legend?, type "Standard Normal Curve" as the Chart Title, and type $z$ and $f(z)$ for Category (X) and Value (Y) titles, respectively. Finally, click Finish.

   **Users of Excel 97/98.** Select cells A2:B73 and click the **ChartWizard**. A dialog box **ChartWizard – Step 1 of 4 – Chart Type** appears.

- In Step 1 select **XY (Scatter)** for Chart Type and the lower right Chart sub-type **Scatter without markers.**
- In Step 2 under the **Data Range** tab the range will already be indicated and the Series radio button for **Columns** will be selected. You may edit the range if it is incorrect. Under the **Series** tab no changes are necessary.
- In Step 3 under the **Titles** tab, type 'Standard Normal Curve' as the Chart Title, $z$ for Value (X) Axis, and $f(z)$ for Value (Y) Axis. Under the **Axes tab**, both check boxes should be selected. Under the **Grid-lines** tab clear all check boxes. Under the **Legend** tab clear the Show legend. Finally, under the **Data Labels** tab select the radio button **None.**
- In Step 4 embed the graph in the current workbook by selecting the radio button **As object in.** Finally, click Finish.

6. Activate the graph for editing and format the display as you wish to present it using the editing features discussed previously.

| | A | B | C | D |
|---|---|---|---|---|
| 1 | z | 0.00 | 0.01 | 0.02 |
| 2 | 0.00 | =NORMSDIST($A2+B$1) | =NORMSDIST($A2+C$1) | =NORMSDIST($A2+D$1) |
| 3 | 0.10 | =NORMSDIST($A3+B$1) | =NORMSDIST($A3+C$1) | =NORMSDIST($A3+D$1) |
| 4 | 0.20 | =NORMSDIST($A4+B$1) | =NORMSDIST($A4+C$1) | =NORMSDIST($A4+D$1) |

Figure 1.34: Formulas For a Normal Table

## Constructing a Normal Table

It is very easy in Excel to produce a table of normal areas. Figure 1.34 shows the formulas behind the workbook in Figure 1.35. Remember that the $ sign prefix makes the corresponding row or column label absolute. This method can be adapted to produce tables of other continuous distributions.

Figure 1.35 gives areas under a standard normal curve for values of $z \geq 0$. It is obtained as follows.

1. Enter the label and values in column A and row 1 as in Figure 1.35. Column A gives the first decimal of $z$ while row 1 gives the second decimal.

2. Enter the formula (Figure 1.35)

$$= \texttt{NORMDIST}(\$A2 + B\$1)$$

in cell B2. Select cell B2, click the fill handle in the lower right corner of B2, and drag to K2.

| | A | B | C | D | E | F | G | H | I | J | K |
|---|---|---|---|---|---|---|---|---|---|---|---|
| 1 | z | **0.00** | **0.01** | **0.02** | **0.03** | **0.04** | **0.05** | **0.06** | **0.07** | **0.08** | **0.09** |
| 2 | **0.00** | 0.5000 | 0.5040 | 0.5080 | 0.5120 | 0.5160 | 0.5199 | 0.5239 | 0.5279 | 0.5319 | 0.5359 |
| 3 | **0.10** | 0.5398 | 0.5438 | 0.5478 | 0.5517 | 0.5557 | 0.5596 | 0.5636 | 0.5675 | 0.5714 | 0.5753 |
| 4 | **0.20** | 0.5793 | 0.5832 | 0.5871 | 0.5910 | 0.5948 | 0.5987 | 0.6026 | 0.6064 | 0.6103 | 0.6141 |
| 5 | **0.30** | 0.6179 | 0.6217 | 0.6255 | 0.6293 | 0.6331 | 0.6368 | 0.6406 | 0.6443 | 0.6480 | 0.6517 |
| 6 | **0.40** | 0.6554 | 0.6591 | 0.6628 | 0.6664 | 0.6700 | 0.6736 | 0.6772 | 0.6808 | 0.6844 | 0.6879 |
| 7 | **0.50** | 0.6915 | 0.6950 | 0.6985 | 0.7019 | 0.7054 | 0.7088 | 0.7123 | 0.7157 | 0.7190 | 0.7224 |
| 8 | **0.60** | 0.7257 | 0.7291 | 0.7324 | 0.7357 | 0.7389 | 0.7422 | 0.7454 | 0.7486 | 0.7517 | 0.7549 |
| 9 | **0.70** | 0.7580 | 0.7611 | 0.7642 | 0.7673 | 0.7704 | 0.7734 | 0.7764 | 0.7794 | 0.7823 | 0.7852 |
| 10 | **0.80** | 0.7881 | 0.7910 | 0.7939 | 0.7967 | 0.7995 | 0.8023 | 0.8051 | 0.8078 | 0.8106 | 0.8133 |
| 11 | **0.90** | 0.8159 | 0.8186 | 0.8212 | 0.8238 | 0.8264 | 0.8289 | 0.8315 | 0.8340 | 0.8365 | 0.8389 |
| 12 | **1.00** | 0.8413 | 0.8438 | 0.8461 | 0.8485 | 0.8508 | 0.8531 | 0.8554 | 0.8577 | 0.8599 | 0.8621 |
| 13 | **1.10** | 0.8643 | 0.8665 | 0.8686 | 0.8708 | 0.8729 | 0.8749 | 0.8770 | 0.8790 | 0.8810 | 0.8830 |
| 14 | **1.20** | 0.8849 | 0.8869 | 0.8888 | 0.8907 | 0.8925 | 0.8944 | 0.8962 | 0.8980 | 0.8997 | 0.9015 |
| 15 | **1.30** | 0.9032 | 0.9049 | 0.9066 | 0.9082 | 0.9099 | 0.9115 | 0.9131 | 0.9147 | 0.9162 | 0.9177 |
| 16 | **1.40** | 0.9192 | 0.9207 | 0.9222 | 0.9236 | 0.9251 | 0.9265 | 0.9279 | 0.9292 | 0.9306 | 0.9319 |
| 17 | **1.50** | 0.9332 | 0.9345 | 0.9357 | 0.9370 | 0.9382 | 0.9394 | 0.9406 | 0.9418 | 0.9429 | 0.9441 |
| 18 | **1.60** | 0.9452 | 0.9463 | 0.9474 | 0.9484 | 0.9495 | 0.9505 | 0.9515 | 0.9525 | 0.9535 | 0.9545 |
| 19 | **1.70** | 0.9554 | 0.9564 | 0.9573 | 0.9582 | 0.9591 | 0.9599 | 0.9608 | 0.9616 | 0.9625 | 0.9633 |
| 20 | **1.80** | 0.9641 | 0.9649 | 0.9656 | 0.9664 | 0.9671 | 0.9678 | 0.9686 | 0.9693 | 0.9699 | 0.9706 |
| 21 | **1.90** | 0.9713 | 0.9719 | 0.9726 | 0.9732 | 0.9738 | 0.9744 | 0.9750 | 0.9756 | 0.9761 | 0.9767 |
| 22 | **2.00** | 0.9772 | 0.9778 | 0.9783 | 0.9788 | 0.9793 | 0.9798 | 0.9803 | 0.9808 | 0.9812 | 0.9817 |
| 23 | **2.10** | 0.9821 | 0.9826 | 0.9830 | 0.9834 | 0.9838 | 0.9842 | 0.9846 | 0.9850 | 0.9854 | 0.9857 |
| 24 | **2.20** | 0.9861 | 0.9864 | 0.9868 | 0.9871 | 0.9875 | 0.9878 | 0.9881 | 0.9884 | 0.9887 | 0.9890 |
| 25 | **2.30** | 0.9893 | 0.9896 | 0.9898 | 0.9901 | 0.9904 | 0.9906 | 0.9909 | 0.9911 | 0.9913 | 0.9916 |
| 26 | **2.40** | 0.9918 | 0.9920 | 0.9922 | 0.9925 | 0.9927 | 0.9929 | 0.9931 | 0.9932 | 0.9934 | 0.9936 |
| 27 | **2.50** | 0.9938 | 0.9940 | 0.9941 | 0.9943 | 0.9945 | 0.9946 | 0.9948 | 0.9949 | 0.9951 | 0.9952 |
| 28 | **2.60** | 0.9953 | 0.9955 | 0.9956 | 0.9957 | 0.9959 | 0.9960 | 0.9961 | 0.9962 | 0.9963 | 0.9964 |
| 29 | **2.70** | 0.9965 | 0.9966 | 0.9967 | 0.9968 | 0.9969 | 0.9970 | 0.9971 | 0.9972 | 0.9973 | 0.9974 |
| 30 | **2.80** | 0.9974 | 0.9975 | 0.9976 | 0.9977 | 0.9977 | 0.9978 | 0.9979 | 0.9979 | 0.9980 | 0.9981 |
| 31 | **2.90** | 0.9981 | 0.9982 | 0.9982 | 0.9983 | 0.9984 | 0.9984 | 0.9985 | 0.9985 | 0.9986 | 0.9986 |
| 32 | **3.00** | 0.9987 | 0.9987 | 0.9987 | 0.9988 | 0.9988 | 0.9989 | 0.9989 | 0.9989 | 0.9990 | 0.9990 |
| 33 | **3.10** | 0.9990 | 0.9991 | 0.9991 | 0.9991 | 0.9992 | 0.9992 | 0.9992 | 0.9992 | 0.9993 | 0.9993 |
| 34 | **3.20** | 0.9993 | 0.9993 | 0.9994 | 0.9994 | 0.9994 | 0.9994 | 0.9994 | 0.9995 | 0.9995 | 0.9995 |
| 35 | **3.30** | 0.9995 | 0.9995 | 0.9995 | 0.9996 | 0.9996 | 0.9996 | 0.9996 | 0.9996 | 0.9996 | 0.9997 |
| 36 | **3.40** | 0.9997 | 0.9997 | 0.9997 | 0.9997 | 0.9997 | 0.9997 | 0.9997 | 0.9997 | 0.9997 | 0.9998 |
| 37 | **3.50** | 0.9998 | 0.9998 | 0.9998 | 0.9998 | 0.9998 | 0.9998 | 0.9998 | 0.9998 | 0.9998 | 0.9998 |
| 38 | **3.60** | 0.9998 | 0.9998 | 0.9999 | 0.9999 | 0.9999 | 0.9999 | 0.9999 | 0.9999 | 0.9999 | 0.9999 |
| 39 | **3.70** | 0.9999 | 0.9999 | 0.9999 | 0.9999 | 0.9999 | 0.9999 | 0.9999 | 0.9999 | 0.9999 | 0.9999 |
| 40 | **3.80** | 0.9999 | 0.9999 | 0.9999 | 0.9999 | 0.9999 | 0.9999 | 0.9999 | 0.9999 | 0.9999 | 0.9999 |
| 41 | **3.90** | 1.0000 | 1.0000 | 1.0000 | 1.0000 | 1.0000 | 1.0000 | 1.0000 | 1.0000 | 1.0000 | 1.0000 |

Figure 1.35: Constructing a Normal Table

3. Select cells B2:K2, click the fill handle in the lower right corner of K2, and drag to K41 to fill the block B2:K41.

## 1.6 Boxplots

Excel does not provide a boxplot. However the Microsoft Personal Support Center has a web page "How to Create a BoxPlot– Box and Whisker Chart" located at

*http://support.microsoft.com/support/kb/articles/q155/1/30.asp*

with instructions for creating a reasonable boxplot in Excel 5/95. We have modified these for use with Excel 97/98 and we illustrate by constructing side-by-side boxplots of the calorie data for beef, meat, and poultry hot dogs in Example 1.3.

| | A | B | C | D | E | F | G |
|---|---|---|---|---|---|---|---|
| 1 | | | *Boxplots of Hot Dog Calories* | | | | |
| 2 | Beef | Meat | Poultry | | Beef | Meat | Poultry |
| 3 | 186 | 173 | 129 | median | 152.5 | 153 | 129 |
| 4 | 181 | 191 | 132 | Q1 | 140.5 | 139 | 102 |
| 5 | 176 | 182 | 102 | min | 111 | 107 | 86 |
| 6 | 149 | 190 | 106 | max | 190 | 195 | 170 |
| 7 | 184 | 172 | 94 | Q3 | 177.25 | 179 | 143 |
| 8 | 190 | 147 | 102 | | | | |
| 9 | 158 | 146 | 87 | | | | |
| 10 | 139 | 139 | 99 | *Formulas for Beef Column* | | | |
| 11 | 175 | 175 | 170 | median | =MEDIAN(A3:A22) | | |
| 12 | 148 | 136 | 113 | Q1 | =QUARTILE(A3:A22,1) | | |
| 13 | 152 | 179 | 135 | min | =MIN(A3:A22) | | |
| 14 | 111 | 153 | 142 | max | =MAX(A3:A22) | | |
| 15 | 141 | 107 | 86 | Q3 | =QUARTILE(A3:A22,3) | | |
| 16 | 153 | 195 | 143 | | | | |
| 17 | 190 | 135 | 152 | | | | |
| 18 | 157 | 140 | 146 | | | | |
| 19 | 131 | 138 | 144 | | | | |
| 20 | 149 | | | | | | |
| 21 | 135 | | | | | | |
| 22 | 132 | | | | | | |

Figure 1.36: Boxplot – Data and Preparation

1. Step 1. Enter the calorie data into three columns of a workbook (Figure 1.36), then find and enter the five-number summary into another three columns **in the order** median, first quartile, minimum, maximum, third quartile. We have entered this information, including labels in block D3:G7 in Figure 1.36.

2. Step 2. Select cells D2:G7 and click on the **ChartWizard** button. Choose the **Stock** Chart type from the selections on the left and then from the Chart sub-types on the right select **Volume-Open-High-Low-Close Chart** (at the bottom right in Figure 1.37). Click Next.

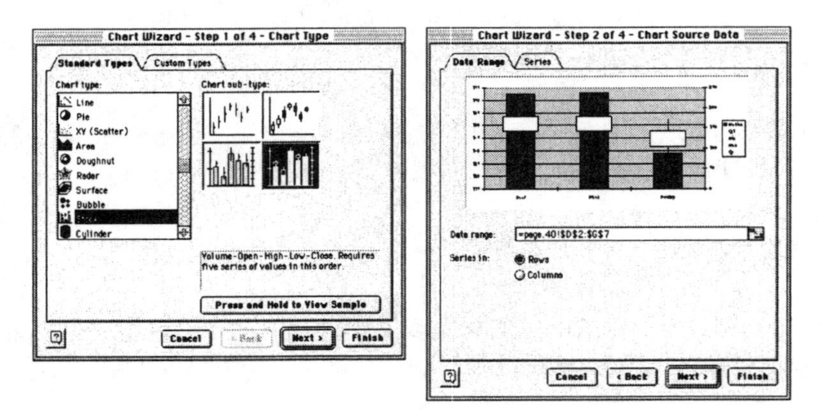


Figure 1.37: Boxplot – ChartWizard Steps 1 and 2

3. Step 3. Check the radio button **Rows** for Series in: and click Next.

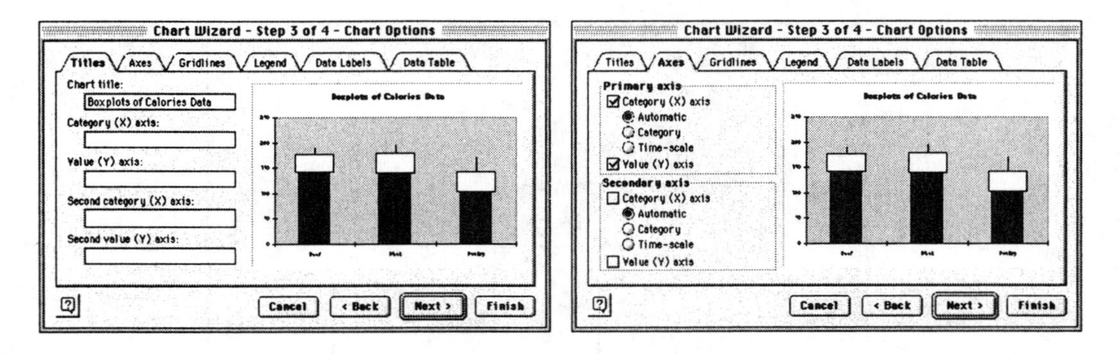


Figure 1.38: Boxplot – ChartWizard Step 3

4. Step 4. In the next dialog box (Figure 1.38), under the **Titles** tab enter "Boxplots of Calories Data" for the Chart title, under the **Axes** tab clear the check box next to Value (Y) Axis under **Secondary Axis**, under the **Gridlines** tab clear all boxes, and finally under the **Legend** tab clear the Show Legend box. Click Next.

5. Step 5. In the final dialog box locate the boxplot on your sheet.

Next we edit this chart.

1. Click the chart to activate it. Click once on any one of the colored columns to select the series. Do not click on the white columns. From the Menu Bar choose **Chart – Chart Type...**, select Chart type **Line** on the left and select Sub chart type **Line** (upper left selection on the right side of the dialog box). A line that connects the three white columns appears in the chart.

2. Click once on the line and from the Menu Bar choose **Format – Selected Data Series....**

3. Under the **Patterns** tab, select **None** for Line and **Custom** for Marker. For the custom marker choose the plus sign from the **Style** list, the color black from the **Foreground** list, **None** from the **Background** list, and 5 for the **Size** (font). Click OK.

4. Double-click the Y axis (or equivalently, click the Y axis once to select it and then from the Menu Bar choose **Format – Selected Axis...**) and in the **Format Axis** dialog box under the **Scale** tab set the Minimum to 80. Click OK. The final boxplot appears on your sheet (Figure 1.39).

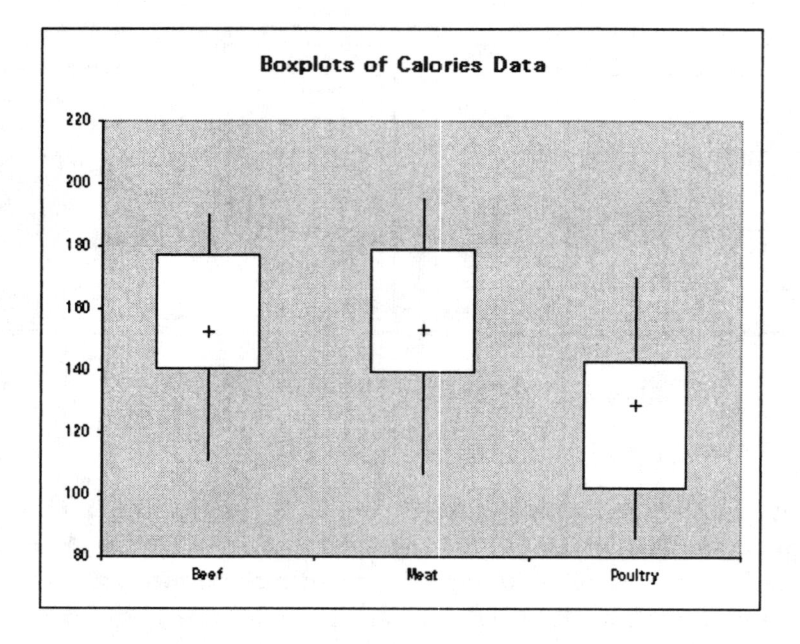


Figure 1.39: Boxplot of Calories Data

These side-by-side boxplots give a vivid graphical summary showing that beef and meat hot dogs are similar in calories and that poultry hot dogs are lower in general.

# Chapter 2

# Examining Relationships

Statistical studies are often carried out to learn whether or how much one measurement (an explanatory variable $x$) can be used to predict the value of another measurement (a response variable $y$). Once data are collected, either through a controlled experiment or an observational study, it is useful to examine graphically if any relationship is justified. We might plot in Cartesian coordinates all pairs $(x_i, y_i)$ of observed values. The resulting graph is called a **Scatterplot**.

## 2.1 Scatterplot

Table 2.1: Manatees Killed and Powerboat Registrations

| Year | 1977 | 1978 | 1979 | 1980 | 1981 | 1982 | 1983 |
|---|---|---|---|---|---|---|---|
| Registrations (1000) | 447 | 460 | 481 | 498 | 513 | 512 | 526 |
| Manatees killed | 13 | 21 | 24 | 16 | 24 | 20 | 15 |

| Year | 1984 | 1985 | 1986 | 1987 | 1988 | 1989 | 1990 |
|---|---|---|---|---|---|---|---|
| Registrations (1000) | 559 | 585 | 614 | 645 | 675 | 711 | 719 |
| Manatees killed | 34 | 33 | 33 | 39 | 43 | 50 | 47 |

**Example 2.1.** (The Endangered Manatee – Exercise 2.4 page 83 in text.) Manatees are large, gentle sea creatures that live along the Florida coast. Many manatees are killed or injured by powerboats. Table 2.1 contains data on powerboat registrations (in thousands) and the number of manatees killed by boats in Florida in the years 1977 to 1990. Make a scatterplot of these data, labeling the axes with the variable names.

## Creating a Scatterplot

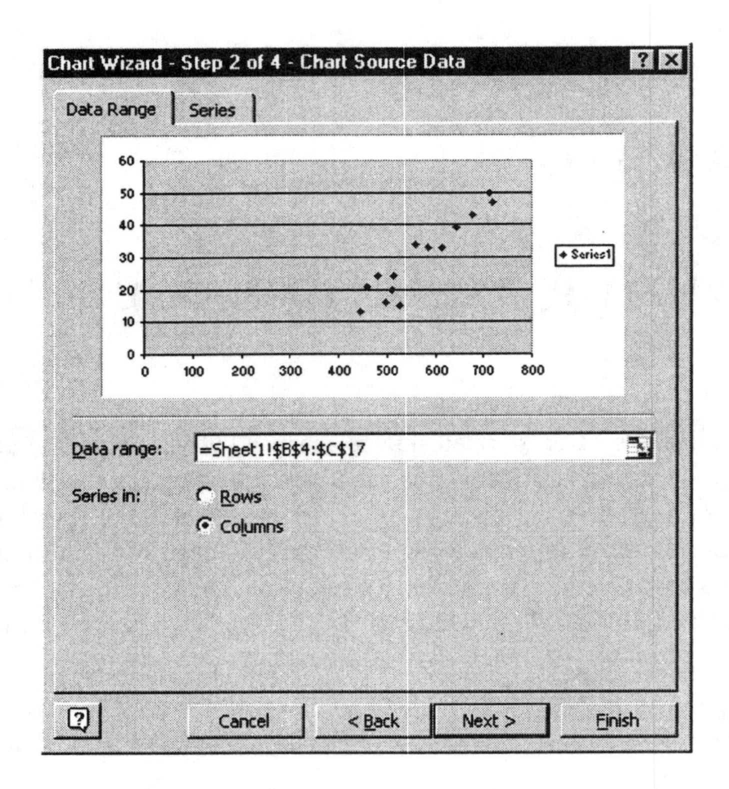


Figure 2.1: Windows ChartWizard – Step 2

The steps involved in creating a scatterplot are similar to those for producing a **Histogram** using the **ChartWizard**. The following instructions are based on **Excel 97/98**. **Excel 5/95** users should instead refer to the Appendix where corresponding instructions are given. Figures 2.1 – 2.4 show the **ChartWizard** interface using Excel 97 on a Toshiba notebook running Windows NT 4.0, rather than on a Macintosh, only for illustration purposes, because a user of Excel should be familiar with the software interface no matter what the operating system.

1. Enter the data from Table 2.1 into cells A4:A17 and C4:C17 of a work book with the labels "Registrations (1000)" and "Manatees killed," referring to Figure 2.4 later in this section.

2. Step 1. Select cells B4:C17 and click on the **ChartWizard**. From the choice of charts select **XY (Scatter)** for Chart type on the left and select the top Chart sub-type **Scatter** on the right. Click Next.

3. Step 2. The next dialog box previews the chart and allows any changes to be made to the data range. Click Next.

4. Step 3. The **Chart Option** dialog box appears (Figure 2.1).
   - Click the **Titles** tab and enter "Powerboat Registrations vs Manatees killed" for Chart title, "Registrations (1000)" for Category (X) axis, and "Manatees killed" for Value (Y) axis.
   - Click the **Legend** tab. Clear the Show Legend check box. Click Next.
5. Step 4. In the last step, select the radio button to enter the chart on the current sheet. Click Finish. The scatterplot appears embedded on your workbook in the selected rectangle (Figure 2.4).

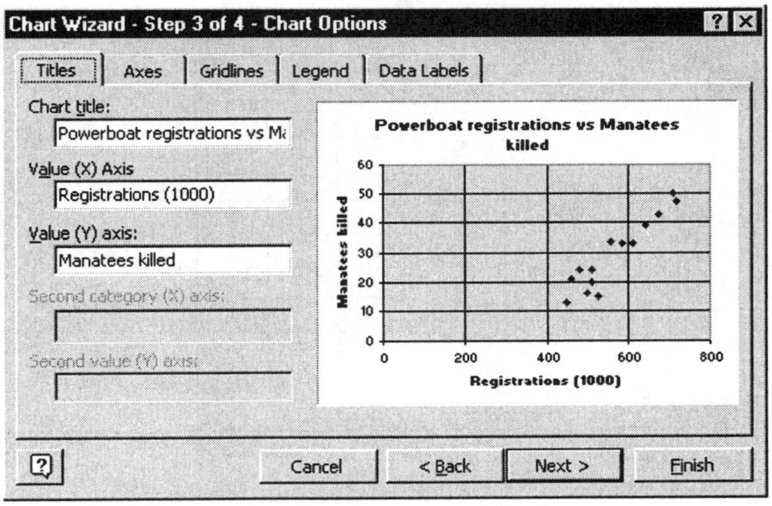

Figure 2.2: Preview of Scatterplot – Step 3

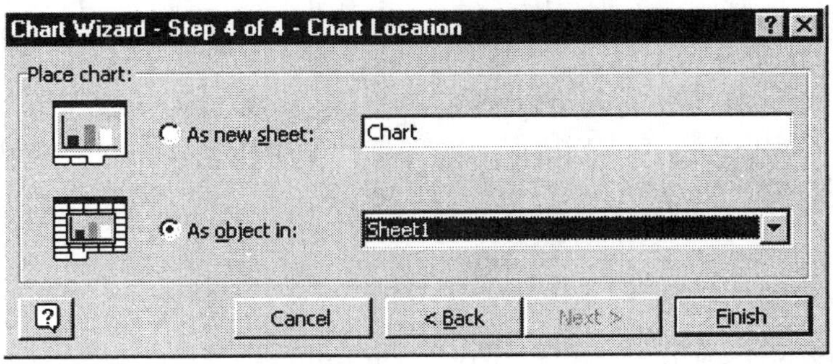

Figure 2.3: Preview of Scatterplot – Step 4

**Note:** Inspection of the scatterplots in Figures 2.2 and 2.4 reveals that the axes scales have been changed. Excel uses a range from 0 to 100% as the default and sometimes the scatterplot will show unwanted blank space, as is the case here. To remedy this, change the horizontal scale. Double click the X axis and in the **Format Axis** dialog box (Figure 1.20) click on the **Scale** to change the zero of the X axis. Similarly select the Y axis for editing. Refer to Figure 1.20 and the discussion for enhancing a histogram in Section 1.2 which applies to any chart, histogram, scatterplot, or other type.

*Powerboat registrations and Manatees killed*

| Year | Registrations (1000) | Manatees killed |
|---|---|---|
| 1977 | 447 | 13 |
| 1978 | 460 | 21 |
| 1979 | 481 | 24 |
| 1980 | 498 | 16 |
| 1981 | 513 | 24 |
| 1982 | 512 | 20 |
| 1983 | 526 | 15 |
| 1984 | 559 | 34 |
| 1985 | 585 | 33 |
| 1986 | 614 | 33 |
| 1987 | 645 | 39 |
| 1988 | 675 | 43 |
| 1989 | 711 | 50 |
| 1990 | 719 | 47 |

Powerboat registrations vs Manatees killed

Figure 2.4: Manatee Data and Scatterplot Embedded on Sheet

## Enhancing a Scatterplot

The scatterplot may be enhanced using editing tools, some of which have been described in Chapter 1. Activate the scatterplot to access new commands which become available under the Menu Bar.

### Labeling a Data Point

By default Excel uses diamonds to plot the points. Suppose, for presentation purposes, we wish to use a different shape (and color) to represent a particular year, 1977, and also to attach the label 1977 to the corresponding point on the scatterplot. The following steps describe how to achieve this.

1. Activate the chart and click on the observation for 1977.

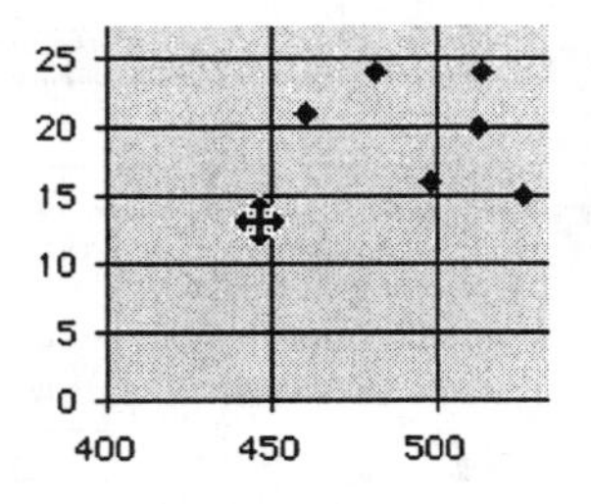


Figure 2.5: Editing a Point

2. Hold down the **Control key (Windows)** or **Command key (Macintosh)** and with your mouse pointer **select** the point representing 1977. The pointer becomes a four-pointed plus sign (Figure 2.5).

3. For **Excel 5/95** choose **Insert – Data Labels** from the Menu Bar to open the **Format Data Point** dialog box. For **Excel 97/98** choose **Format – Selected Data Point...** from the Menu Bar to open a corresponding **Format Data Point** dialog box. Under the **Data Labels** tab select the radio button for Show Value. Click OK. Excel attaches the $y$ value 13 to the point for 1977 on the scatterplot and encloses it within a grey bordered selection box for editing. Type "1977" (which appears in the **Formula Bar**) and press enter. The selection box now contains the year 1977. Move it to a convenient place and **deselect** by clicking elsewhere.

## Changing the Marker and Color of a Data Point

To make the year 1977 stand out more, we will now change the symbol and color of its point on the scatterplot.

1. Activate the chart and click on the observation for 1977.

2. Hold down the **Control key (Windows)** or **Command key (Macintosh)** and select the point representing 1977. The pointer becomes a four-pointed plus sign (Figure 2.5).

3. For **Excel 5/95** choose **Insert – Data Labels** from the Menu Bar to open the **Format Data Point** dialog box. For **Excel 97/98** choose **Format – Selected Data Point...** from the Menu Bar to open a corresponding **Format Data Point** dialog box. Under the **Patterns** tab, leave the **Line** selection as **None.** Under **Marker** select a marker type from the pull-down list for **Style**, and also select a Foreground and Background color (Figure 2.6). In **Excel 5/95** the size of the marker cannot be changed and there is no **Options** tab. Your selection is previewed in the small **Sample** box in

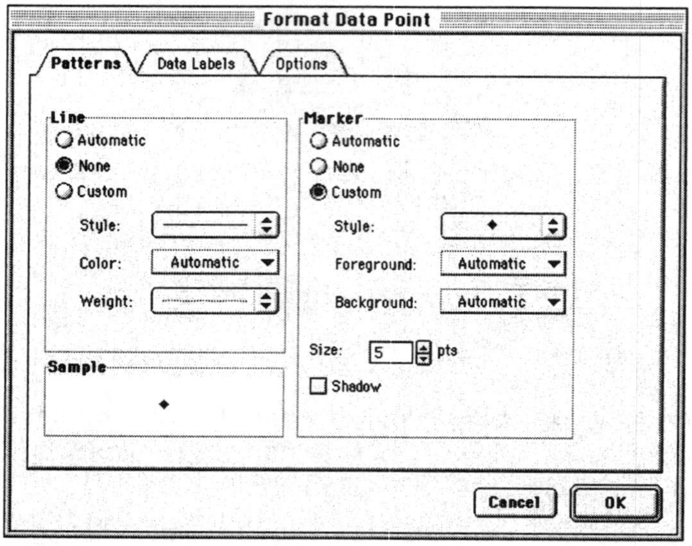

Figure 2.6: Changing the Default Marker

the lower portion of the dialog box. Click OK. The final scatterplot appears in Figure 2.7.

## 2.2 Correlation

The correlation between two variables $x$ and $y$ measures the strength of the linear association between them. For $n$ pairs $(x_i, y_i)$, $1 \leq i \leq n$, of data points the sample correlation coefficient is defined to be

$$r = \frac{1}{n-1} \sum_{i=1}^{n} \left( \frac{x_i - \bar{x}}{s_x} \right) \left( \frac{y_i - \bar{y}}{s_y} \right)$$

where $\bar{x}$ and $\bar{y}$ are the sample means of the $\{x_i\}$ and $\{y_i\}$, respectively, and $s_x$ and $s_y$ are the corresponding sample standard deviations.

### Using the CORREL Function

The most direct way to find the correlation for Example 2.1 is by the **Function Wizard** for **Excel 5/95** or the **Formula Palette** for **Excel 97/98** to describe the function `CORREL` which computes the correlation coefficient.

**Example 2.2.** Find the correlation between powerboat registrations and manatees killed for the data set in Example 2.1.

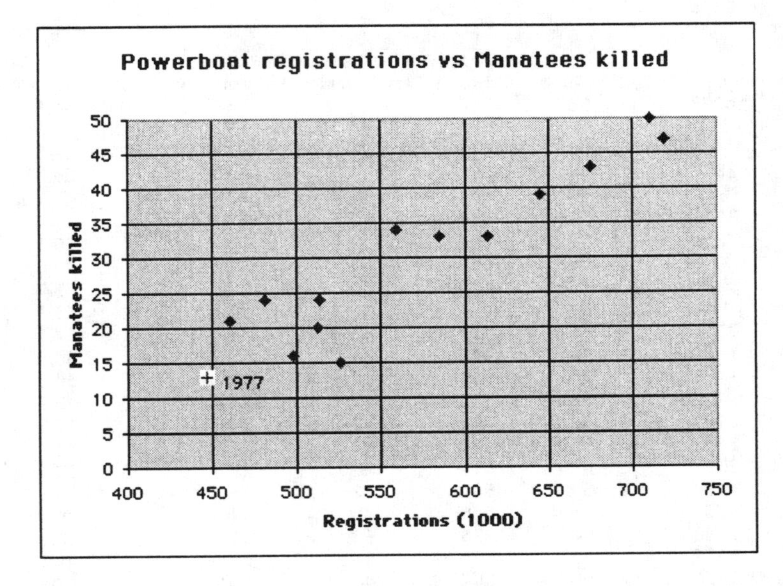


Figure 2.7: Scatterplot with Labelled Point

**Solution.**

1. Open the workbook containing the data and labels you used in Example 2.1. (Refer to Figure 2.4.) Select an empty cell where you want the correlation to appear and invoke either the **Function Wizard** or the **Formula Palette**. In each case, select **Statistical** for Function Category and `CORREL` for Function Name.

2. Enter B4:B17 for **Array1** and C4:C17 for **Array2**. You may enter by hand or click and drag on the workbook over the range B4:B17, press tab on the keyboard, then click and drag over the range B2:B21, and finally click OK (or Finish, for **Excel 5/95**). The answer 0.9415 appears in the cell you selected.

## Using the ToolPak

Correlation between two variables can also be calculated using the **Correlation** tool in the **Analysis ToolPak**. This tool is most effective, however, for determining pairwise correlations for multivariate data sets for which repeated use of the `CORREL` function would be inefficient.

This tool prints out a matrix of correlations. Such a matrix is helpful in multiple regression in deciding which variables to include in a model.

**Example 2.3.** (EESEE Data set from the CD. Source: *Statistical Abstract of the United States.* See also Table 1.6 page 26 and Example 2.3 page 82 in text.) Figure 2.8 shows the first 25 sets of observations.

| | A | B | C | D | E | F |
|---|---|---|---|---|---|---|
| 1 | *Education and related data for the states* | | | | | |
| 2 | Population (1000) | SAT verbal | SAT math | Percent taking | Percent no HS | Teachers' pay ($1000) |
| 3 | 4273 | 565 | 558 | 8 | 33.1 | 31.3 |
| 4 | 607 | 521 | 513 | 47 | 13.4 | 49.6 |
| 5 | 4428 | 525 | 521 | 28 | 21.3 | 32.5 |
| 6 | 2510 | 566 | 550 | 6 | 33.7 | 29.3 |
| 7 | 31878 | 495 | 511 | 45 | 23.8 | 43.1 |
| 8 | 3823 | 536 | 538 | 30 | 15.6 | 35.4 |
| 9 | 3274 | 507 | 504 | 79 | 20.8 | 50.3 |
| 10 | 725 | 508 | 495 | 66 | 22.5 | 40.5 |
| 11 | 543 | 489 | 473 | 50 | 26.9 | 43.7 |
| 12 | 14400 | 498 | 496 | 48 | 25.6 | 33.3 |
| 13 | 7353 | 484 | 477 | 63 | 29.1 | 34.1 |
| 14 | 1184 | 485 | 510 | 54 | 19.9 | 35.8 |
| 15 | 1189 | 543 | 536 | 15 | 20.3 | 30.9 |
| 16 | 11847 | 564 | 575 | 14 | 23.8 | 40.9 |
| 17 | 5841 | 494 | 494 | 57 | 24.4 | 37.7 |
| 18 | 2852 | 590 | 600 | 5 | 19.9 | 32.4 |
| 19 | 2572 | 579 | 571 | 9 | 18.7 | 35.1 |
| 20 | 3884 | 549 | 544 | 12 | 35.4 | 33.1 |
| 21 | 4351 | 559 | 550 | 9 | 31.7 | 26.8 |
| 22 | 1243 | 504 | 498 | 68 | 21.2 | 32.9 |
| 23 | 5072 | 507 | 504 | 64 | 21.6 | 41.2 |
| 24 | 6092 | 507 | 504 | 80 | 20 | 42.9 |
| 25 | 9594 | 557 | 565 | 11 | 23.2 | 44.8 |
| 26 | 4658 | 582 | 593 | 9 | 17.6 | 36.9 |
| 27 | 2716 | 569 | 557 | 4 | 35.7 | 27.7 |

Figure 2.8: Education and Related Data – EESEE Data

These are data about the individual states which pertain to education: for instance, "SAT math" is the average SAT math score of the states' high school seniors, "Percent taking" is the percent of seniors in the state who take the SAT, and so on. Find the correlations between each pair of variables.

**Solution.**

1. From the Menu Bar choose **Tools – Data Analysis** and in the dialog box highlight **Correlation** and click OK.

2. In the next dialog box **Correlation**, enter A2:F27 for **Input range** (most conveniently done by clicking and dragging over this range on the workbook and pressing the Tab key). Check the box **Labels in first row** and point to cell G1 for **Output range**. Click OK.

### Excel Output

The output appears in G1:M7 as shown in Figure 2.9. From cell I4 we read that the correlation between SAT math and SAT verbal is 0.9580; from cell J5 we read that the correlation between SAT math and the percent of seniors taking the SAT is −0.8583; and so on. In view of symmetry only half the correlation matrix is required.

| | G | H | I | J | K | L | M |
|---|---|---|---|---|---|---|---|
| 1 | | *Population* | *SAT verbal* | *SAT math* | *Percent taking* | *Percent no HS* | *Teachers' pay* |
| 2 | Population | 1 | | | | | |
| 3 | SAT verbal | -0.1880 | 1 | | | | |
| 4 | SAT math | -0.0505 | 0.9580 | 1 | | | |
| 5 | Percent taking | 0.0422 | -0.8835 | -0.8583 | 1 | | |
| 6 | Percent no HS | 0.0620 | 0.1612 | 0.0427 | -0.3349 | 1 | |
| 7 | Teachers' pay | 0.1890 | -0.4053 | -0.3345 | 0.5514 | -0.5230 | 1 |

Figure 2.9: Correlation ToolPak Output

## 2.3 Least-Squares Regression

We have seen how to plot two variables against each other in a scatterplot and have calculated the correlation coefficient to measure the strength of the linear association between them. It is useful to have an analytic relationship between the explanatory variable $x$ and the response variable $y$ of the form

$$y = f(x)$$

for predicting $y$ from $x$. Such a relationship is called a simple (meaning one explanatory variable) **regression curve.** The simplest curve is a straight line

$$y = a + bx$$

called the regression line of $y$ on $x$. The regression line represents, under certain assumptions, the mean response at each specified value $x$.

The method used to determine the coefficients $a$ and $b$ goes back at least to the great mathematician Gauss and is called the **Principle of Least-Squares.** Gauss himself recognized that the criterion was arbitrary and he used it because the coefficients $a$ and $b$ were then solvable in closed form. (Additonal reasons connected with the errors being normal are presented in more advanced treatments.)

For a given $x_i$ we call

$$\hat{y}_i = a + bx_i$$

the **predicted** value and

$$e_i = y_i - \hat{y}_i$$

is called the **residual.** The **error sum of squares** is defined to be

$$\sum_{i=1}^{n} e_i^2 = \sum_{i=1}^{n} (y_i - a - bx_i)^2.$$

By differentiating with respect to $a$ and $b$ we can solve for the values which minimize $\sum_{i=1}^{n} e_i^2$. These are the values used in the regression line. They are given by the formulas

$$\begin{aligned} \text{slope} \quad b &= r\,\frac{s_y}{s_x} \\ \text{intercept} \quad a &= \bar{y} - b\bar{x} \end{aligned}$$

where $\bar{x} = \frac{1}{n}\sum_{i=1}^{n} x_i$, $\bar{y} = \frac{1}{n}\sum_{i=1}^{n} y_i$, $r$ is the correlation coefficient, and

$$(n-1)s_x^2 = \sum_{i=1}^{n}(x_i - \bar{x})^2 = \sum_{i=1}^{n} x_i^2 - \frac{1}{n}\left(\sum_{i=1}^{n} x_i\right)^2$$

$$(n-1)s_y^2 = \sum_{i=1}^{n}(y_i - \bar{y})^2 = \sum_{i=1}^{n} y_i^2 - \frac{1}{n}\left(\sum_{i=1}^{n} y_i\right)^2$$

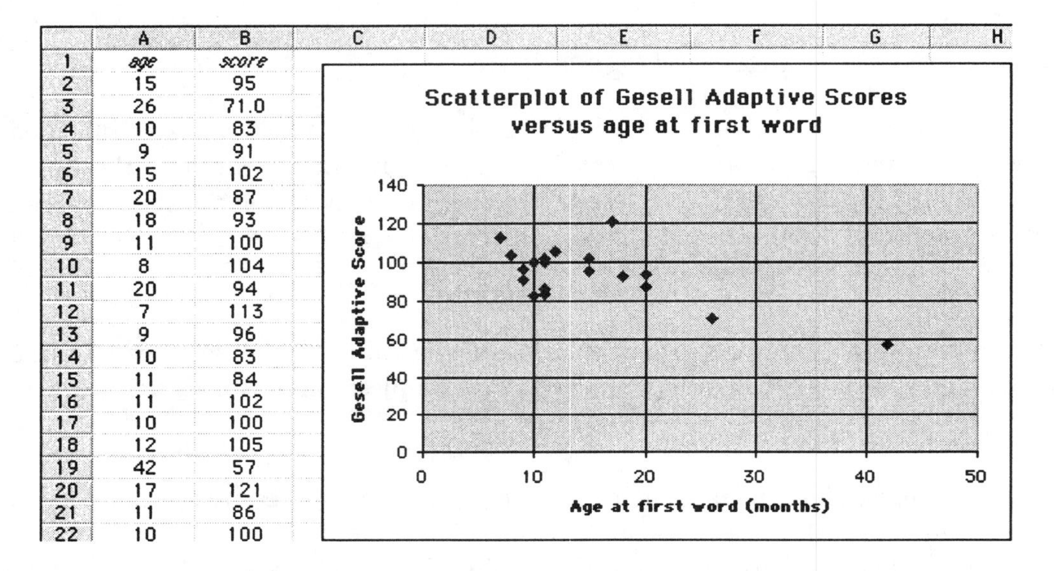

| | A | B |
|---|---|---|
| 1 | age | score |
| 2 | 15 | 95 |
| 3 | 26 | 71.0 |
| 4 | 10 | 83 |
| 5 | 9 | 91 |
| 6 | 15 | 102 |
| 7 | 20 | 87 |
| 8 | 18 | 93 |
| 9 | 11 | 100 |
| 10 | 8 | 104 |
| 11 | 20 | 94 |
| 12 | 7 | 113 |
| 13 | 9 | 96 |
| 14 | 10 | 83 |
| 15 | 11 | 84 |
| 16 | 11 | 102 |
| 17 | 10 | 100 |
| 18 | 12 | 105 |
| 19 | 42 | 57 |
| 20 | 17 | 121 |
| 21 | 11 | 86 |
| 22 | 10 | 100 |



Figure 2.10: Gesell Data and Scatterplot

## Determining the Regression Line

Excel provides three built-in methods for regression analysis: **Trendline**, the **Regression** tool in the **Analysis ToolPak**, and regression functions, such `FORECAST` and `TREND`. For merely graphing a regression line, and providing its equation and the coefficient of determination $r^2$, the **Trendline** command suffices. Later, in dealing with regression in more detail, we will develop use of the **Regression** tool as well as regression functions.

### Linear Trendline

We use the **Linear Trendline** to insert a curve on a scatterplot. The trendline can be added to any scatterplot even after the **Regression** tool is used.

**Example 2.4.** (Example 2.12 page 117 in text.) Does the age at which a child begins to talk predict later scores on a test of mental ability? A study of the development of young children recorded the age in months at which each of 21 children spoke their first word and their Gesell Adaptive Score, the result of an aptitude test taken much later. The data appear in Table 2.6 in the text and are given in Figure 2.10. Construct a scatterplot and superimpose the least-squares regression line of response variable, $y$ = score, against the explanatory variable, $x$ = age. Fit the least-square regression line to the data.

**Solution.** We first construct a scatterplot of the data (also shown in Figure 2.10) to verify that a linear model is appropriate.

1. Enter the data in cells A2:B22 of a workbook. Create a scatterplot with the **ChartWizard** and edit it so it appears as shown in Figure 2.10. The scatterplot shows an approximate linear relationship with a negative. It is therefore appropriate to fit the data pairs with a straight line as a first approximation.

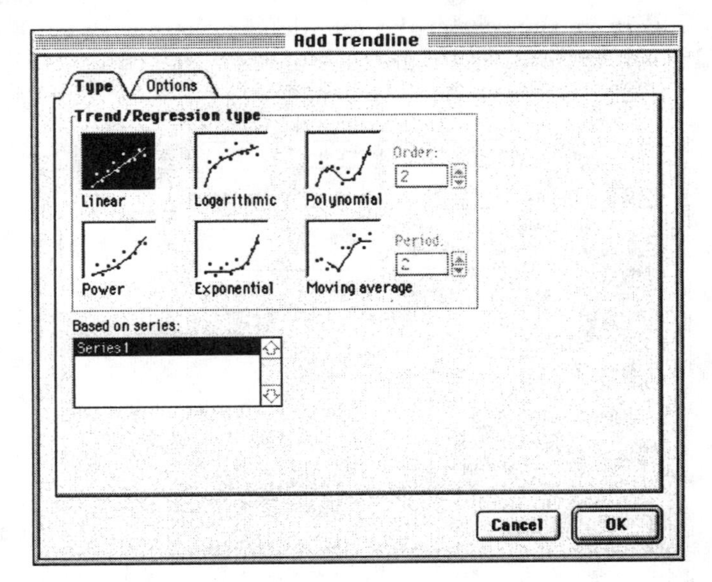


Figure 2.11: Trendline Type

2. Activate the chart for editing and select the data by **clicking on one of the points**. The points appear highlighted, the **Name box** in the **Formula Bar** shows **S1** and in the text entry area we can read

$$= \texttt{SERIES}(,\text{Sheet1!\$}B\text{\$1}, \text{Sheet1!\$}A\text{\$2} : \text{\$}A\text{\$22}, \ \text{Sheet1!\$}B\text{\$2} : \text{\$}B\text{\$22}, 1)$$

meaning that the series has been selected. (Refer to the online help for more information on this function and the Introduction for the meaning of Sheet1! notation.)

3. For **Excel 5/95** choose **Insert – Trendline** from the Menu Bar, while for **Excel 97/98** choose **Chart – Add Trendline...** from the Menu Bar. Then proceed as follows. Click the **Type** tab and select **Linear** (Figure 2.11). **Excel 97/98** has an additional text area (**Based on series**) in the blank space at the bottom of this figure. Click the **Options** tab and select the radio button **Automatic:Linear (Series1)** (Figure 2.12). Check the boxes **Display Equation on Chart** and **Display R-squared Value on Chart** Make sure that **Set Intercept** box is clear. Click OK. The regression line is superimposed on the scatterplot, its equation displayed $y = -1.127x+109.87$, and the coefficient of determination $r^2 = 0.41$ is inserted on the scatterplot.

4. Edit the graph for presentation purposes. Activate the chart and and click on the **rectangular box** surrounding the equation; the border turns a thicker grey color. Use the Decimal tool to increase the number of decimal points. Edit the text by replacing $x$ with "age" and $y$ with "score." Move $r^2 = 0.4100$ from its location on the graph to a more convenient place. Also label Child 18 and Child 19 on the scatterplot and denote them with a different marker. The final result appears as Figure 2.13.

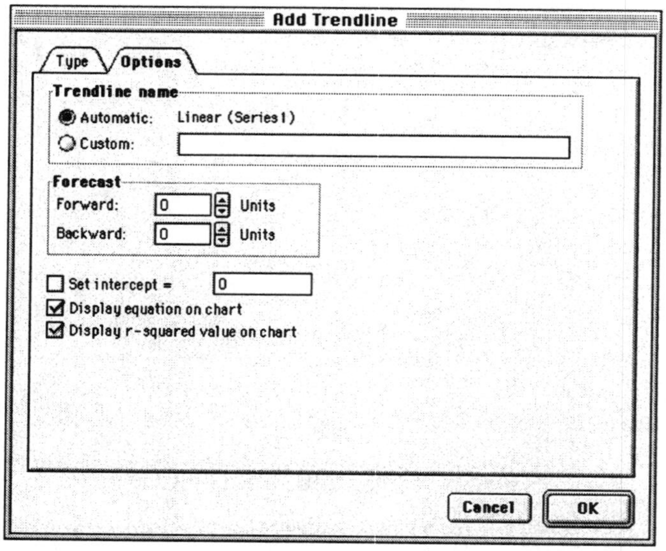

Figure 2.12: Trendline Option

# Residuals

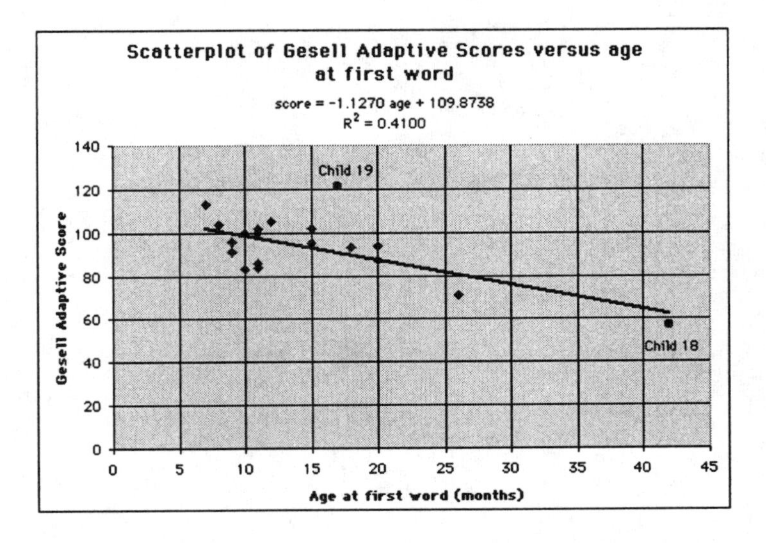


Figure 2.13: Regression Line and Scatterplot

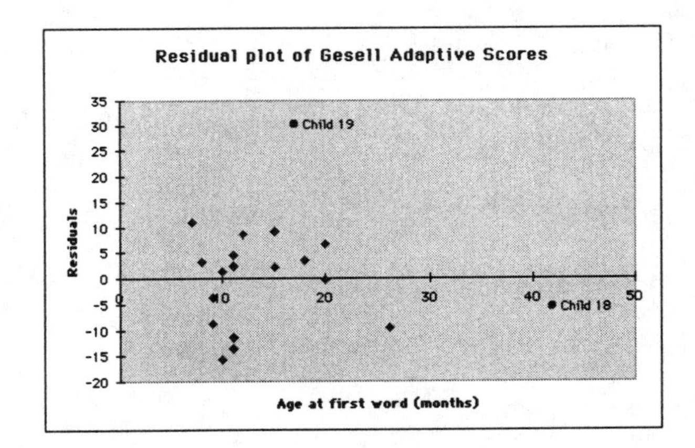


Figure 2.14: Residual Plot

Residuals provide evidence of how well the regression model fits, and are examined in Chapter 11 when the **Regression** tool is introduced. Figure 2.14 shows a scatterplot, using the Regression tool, of residuals against age for the Gesell scores. Child 18 and Child 19 have been identified on the residual plot. Child 19 is considered an *outlier*, an observation which lies outside the overall pattern of fit (in the $y$ direction) of the other observations. Child 18 is an *influential* point because it has pulled the regression line close to it.

# Chapter 3

# Producing Data

A sample is a subset of a population on which measurements are made and which the statistician uses to draw conclusions (inferences) about the entire population. Excel provides tools for sampling from a specified population, for calculating probabilities associated with the standard models and their inverse cumulative distributions using the Excel **functions**, and for simulating values from probability distributions using both the `RAND()` function and the **Random Number Generation** tool.

## 3.1 Simple Random Samples (SRS)

The Excel function `RAND()` picks a number uniformly on the interval (0, 1). We can repeatedly select random uniform (0, 1) numbers and assign them to members of a population. Then by sorting the random numbers we obtain a random permutation of the population which provides an SRS of any desired size.

| | A | B |
|---|---|---|
| 1 | *Clients* | *Sample* |
| 2 | 1 | 0.19983942 |
| 3 | 2 | 0.95647962 |
| 4 | 3 | 0.79767353 |
| 5 | 4 | 0.16306893 |
| 6 | 5 | 0.71132734 |
| 7 | 6 | 0.15713574 |
| 8 | 7 | 0.44142466 |
| 9 | 8 | 0.44292645 |
| 10 | 9 | 0.19194526 |
| 11 | 10 | 0.30569203 |
| 12 | 11 | 0.41276885 |

Figure 3.1: Simple Random Sample

**Example 3.1.** (Example 3.4 page 172 in text.) Joan's small accounting firm serves 30 business clients. Joan wants to interview a sample of 5 clients in detail to find ways to improve client satisfaction.

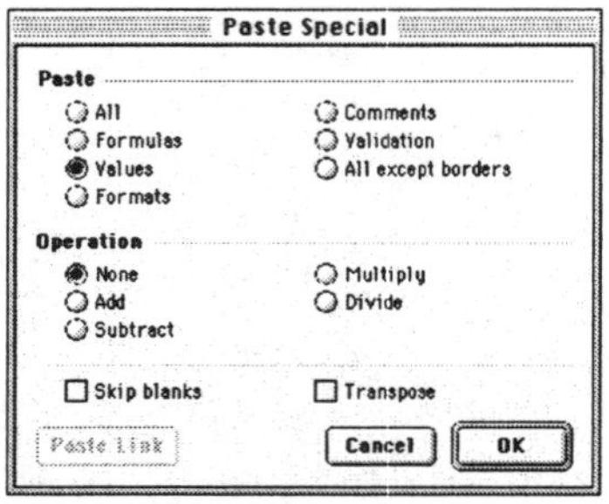

Figure 3.2: Paste Special Dialog Box

Use Excel to find a simple random sample of size 5 from a population of size 30.

**Solution.**

1. Give each client a unique numerical label from the set $\{1, 2, 3, \ldots, 30\}$ and enter the values in cells A2:A31 of a workbook. Enter the label "Clients" in cell A1 and "Sample" in cell B1.

2. Enter `= RAND()` in cell B2 and fill down to B31. The function `RAND()` selects a number uniformly in (0, 1). Figure 3.1 shows a portion of the workbook.

3. Select cells B2:B31 and from the Menu Bar choose **Edit – Copy**. Then, with B2:B31 **still selected**, choose **Edit – Paste Special** from the Menu Bar. (**Windows** users can click the **right mouse button** while **Macintosh** users should hold down the **Option – Command** keys and click to get the **Shortcut Menu** box.) Select **Paste – Special** and in the **Paste – Special** dialog box select the radio buttons for **Values** and **None** (Figure 3.2). This has the effect of replacing the formulas in the cells of column B by the actual values they take.

4. Select cells A1:B31, from the Menu Bar choose **Data – Sort**, and in the Sort By drop down list, click the arrow, and select **Sample**. Also select the radio button for **Header Row** (Figure 3.3). Excel sorts the data in ascending order in column B and carries the order to column A which gives a random permutation of column A.

5. Use the numerical labels in cells A2:A6 to designate the clients in the SRS in the control group. Figure 3.4 shows a portion of the result of the sort. We read off the set of clients {12, 18, 23, 1, 6} for the SRS.

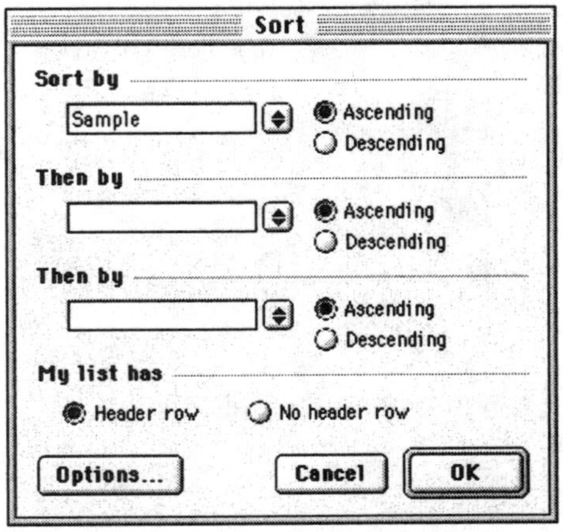

Sort
Sort by
Sample
Ascending
Descending
Then by
Ascending
Descending
Then by
Ascending
Descending
My list has
Header row
No header row
Options...
Cancel
OK

Figure 3.3: Sort Dialog Box

| | A | B |
|---|---|---|
| 1 | *Clients* | *Sample* |
| 2 | 12 | 0.01416711 |
| 3 | 18 | 0.04055743 |
| 4 | 23 | 0.07178864 |
| 5 | 1 | 0.10083942 |
| 6 | 6 | 0.15713574 |
| 7 | 4 | 0.16306893 |
| 8 | 27 | 0.16550820 |
| 9 | 9 | 0.19194526 |
| 10 | 24 | 0.24756783 |
| 11 | 10 | 0.30569203 |
| 12 | 20 | 0.30765618 |

Figure 3.4: Sorted Sample

## 3.2 Random Digits

## Using the Sampling Tool

**Example 3.2.** (Exercise 3.13(b) page 179 in text.) It is usual in telephone surveys to use random digit dialing equipment that selects the last four digits of a telephone number at random after being given the exchange (the first three digits). Obtain a random four-digit telephone number.

**Solution.** Each four-digit number is a sample of size four from $\{0, 1, \ldots, 9\}$ taken with replacement. To obtain such a sample:

1. Enter the values $\{0, 1, \ldots, 9\}$ in A2:A11 (Figure 3.5).
2. From the Menu Bar choose **Tools** – **Data Analysis** and select **Sampling** from the dialog box. Click OK.

| | A | B |
|---|---|---|
| 1 | *Digits* | *Sample* |
| 2 | 0 | 7 |
| 3 | 1 | 5 |
| 4 | 2 | 4 |
| 5 | 3 | 3 |
| 6 | 4 | |
| 7 | 5 | |
| 8 | 6 | |
| 9 | 7 | |
| 10 | 8 | |
| 11 | 9 | |

Figure 3.5: Samples with Replacement

3. Complete the **Sampling** dialog box as shown in Figure 3.6 and click **OK**. A random sample of size four appears in cells B2:B5.

Sampling

Input
Input range: $A$1:$A$11
☑ Labels

Sampling Method
○ Periodic
Period:
● Random
Number of samples: 4

Output Options
● Output range: B2
○ New worksheet ply:
○ New workbook

OK
Cancel
Help

Figure 3.6: Samples with Replacement Dialog Box

## Using the RAND() Function

There is another way to generate random digits which we now illustrate by constructing a table of random digits. This is a list of the digits $\{0, 1, \ldots, 9\}$ which has the following properties:

1. The digit in any position in the list has the same chance of being any one of $\{0, 1, \ldots, 9\}$.
2. The digits in different positions are independent in the sense that the value of one has no influence on the value of any other.

You can imagine asking an assistant (or a computer) to mix the digits $\{0, 1, \ldots, 9\}$ in a hat, draw one, then replace the digit drawn, mix again, draw a second digit, and so on.

The Excel function `INT` truncates a real number to its integer value. For instance, = `INT(3.82)` produces the value 3. By combining `INT` and `RAND()` as, for instance

= `INT(10*RAND())`

we can produce random digits from $\{0, 1, \ldots, 9\}$.

**Example 3.3.** Produce a table of 500 random digits.

| | A | B | C | D | E | F | G | H | I | J | K | L | M | N | O | P | Q | R | S | T | U |
|---|---|---|---|---|---|---|---|---|---|---|---|---|---|---|---|---|---|---|---|---|---|
| 1 | | | | | | | | | | *Table of Random Digits* | | | | | | | | | | | |
| 2 | | | | | | | | | | =INT(10*(RAND())) | | | | | | | | | | | |
| 3 | | | | | | | | | | | | | | | | | | | | | |
| 4 | 0 | 4 | 5 | 4 | 2 | 3 | 4 | 4 | 7 | 6 | 3 | 5 | 9 | 7 | 7 | 4 | 5 | 2 | 0 | 7 | |
| 5 | 6 | 2 | 9 | 7 | 0 | 9 | 3 | 5 | 2 | 6 | 9 | 8 | 6 | 1 | 4 | 2 | 7 | 9 | 8 | 2 | |
| 6 | 7 | 8 | 9 | 1 | 5 | 4 | 2 | 0 | 7 | 5 | 2 | 4 | 3 | 3 | 2 | 1 | 7 | 1 | 3 | 8 | |
| 7 | 9 | 5 | 7 | 8 | 9 | 2 | 6 | 3 | 4 | 3 | 6 | 6 | 5 | 0 | 5 | 3 | 0 | 9 | 1 | 8 | |
| 8 | 3 | 4 | 9 | 2 | 5 | 1 | 2 | 4 | 2 | 1 | 2 | 0 | 4 | 1 | 4 | 5 | 3 | 4 | 9 | 8 | |
| 9 | 4 | 6 | 8 | 7 | 6 | 9 | 2 | 8 | 8 | 2 | 0 | 6 | 5 | 5 | 1 | 4 | 1 | 5 | 1 | 6 | |
| 10 | 0 | 5 | 0 | 7 | 4 | 3 | 5 | 8 | 9 | 7 | 9 | 1 | 9 | 3 | 3 | 5 | 3 | 7 | 5 | 2 | |
| 11 | 1 | 3 | 0 | 5 | 1 | 6 | 7 | 2 | 0 | 9 | 3 | 8 | 3 | 5 | 8 | 5 | 3 | 9 | 8 | 6 | |
| 12 | 6 | 3 | 4 | 7 | 2 | 4 | 4 | 6 | 1 | 7 | 0 | 3 | 4 | 4 | 6 | 7 | 2 | 2 | 0 | 1 | |
| 13 | 8 | 3 | 5 | 1 | 9 | 1 | 1 | 0 | 3 | 2 | 9 | 9 | 4 | 9 | 2 | 4 | 7 | 5 | 4 | 5 | |
| 14 | 7 | 4 | 4 | 3 | 4 | 4 | 3 | 1 | 4 | 0 | 6 | 4 | 5 | 0 | 5 | 9 | 0 | 3 | 2 | 7 | |
| 15 | 1 | 1 | 4 | 1 | 1 | 2 | 8 | 2 | 4 | 6 | 6 | 5 | 4 | 8 | 8 | 4 | 2 | 4 | 4 | 8 | |
| 16 | 8 | 7 | 3 | 3 | 2 | 0 | 0 | 8 | 8 | 8 | 3 | 8 | 4 | 5 | 6 | 7 | 3 | 3 | 2 | 9 | |
| 17 | 0 | 4 | 5 | 5 | 5 | 8 | 6 | 6 | 6 | 0 | 6 | 4 | 8 | 0 | 5 | 5 | 3 | 9 | 7 | 3 | |
| 18 | 1 | 6 | 3 | 3 | 7 | 9 | 8 | 9 | 1 | 1 | 6 | 5 | 7 | 9 | 2 | 1 | 1 | 0 | 4 | 1 | |
| 19 | 2 | 8 | 1 | 4 | 4 | 5 | 6 | 5 | 7 | 3 | 6 | 4 | 2 | 4 | 9 | 3 | 6 | 8 | 0 | 7 | |
| 20 | 0 | 5 | 0 | 8 | 7 | 1 | 1 | 6 | 8 | 4 | 0 | 1 | 4 | 2 | 5 | 2 | 0 | 2 | 8 | 4 | |
| 21 | 7 | 0 | 2 | 5 | 1 | 6 | 7 | 0 | 8 | 3 | 4 | 9 | 5 | 9 | 9 | 2 | 8 | 5 | 0 | 8 | |
| 22 | 1 | 7 | 8 | 9 | 8 | 1 | 8 | 3 | 5 | 3 | 4 | 5 | 0 | 3 | 3 | 7 | 1 | 7 | 1 | 6 | |
| 23 | 4 | 0 | 5 | 7 | 6 | 7 | 0 | 3 | 2 | 7 | 6 | 1 | 1 | 8 | 7 | 5 | 5 | 6 | 4 | 3 | |
| 24 | 4 | 2 | 5 | 9 | 3 | 4 | 0 | 4 | 5 | 4 | 8 | 9 | 6 | 2 | 2 | 4 | 8 | 6 | 1 | 6 | |
| 25 | 6 | 2 | 6 | 1 | 9 | 2 | 8 | 8 | 2 | 5 | 4 | 3 | 4 | 1 | 8 | 9 | 4 | 7 | 8 | 6 | |
| 26 | 1 | 0 | 4 | 1 | 6 | 5 | 7 | 5 | 1 | 8 | 5 | 7 | 0 | 9 | 0 | 2 | 7 | 3 | 9 | 0 | |
| 27 | 1 | 2 | 0 | 4 | 1 | 6 | 8 | 7 | 6 | 3 | 7 | 4 | 9 | 1 | 4 | 2 | 4 | 4 | 2 | 9 | |
| 28 | 8 | 2 | 8 | 0 | 7 | 6 | 0 | 1 | 5 | 5 | 9 | 8 | 4 | 8 | 9 | 9 | 4 | 1 | 9 | 4 | |

Figure 3.7: Table of Random Digits

**Solution.**

1. In cell A4 of a workbook enter the formula = `INT(10* RAND())`.
2. Select cell A4, click the fill handle in the lower right corner and drag across to cell T4.
3. Select cells A4:T4, click the fill handle in the lower right corner and drag down to cell T28.

The result is shown in Figure 3.7. Your digits will necessarily be different as they are random. Press the **F9** key and see what happens. The entire table changes as new random digits are now **dynamically selected**.

# Chapter 4

# Probability and Sampling Distributions

A phenomenon is called **random** if individual outcomes are uncertain but there is still a regular distribution of outcomes in a large number of repetitions. The **probability** of any outcome of a random phenomenon is the proportion of times the outcome would occur in a very long series of repetitions. One way to develop an intuition for randomness is to observe random behavior.

## 4.1 Randomness

A real world probability can only be estimated by observing data. Computer simulations are useful because they help develop insight into the meaning of random variation. Excel is well-suited for simulation and provides both a `RAND()` function and a **Random Number Generation** tool for such purpose.

### Tossing a Fair Coin

> **Example 4.1.** (Example 4.2 page 216 in text.) When you toss a coin, there are only two possible outcomes, heads or tails. Using the `RAND()` function simulate 1000 independent tosses of a fair coin and plot on a graph the proportion of heads after each toss. Also show on the same graph a horizontal line at the height 0.5.

**Solution.** The `RAND()` function produces a number uniformly distributed on the interval (0, 1). This can be converted into integers taking the values 0 or 1 with equal probability if this uniform random number is multiplied by 2 and then the integer part is taken. The Excel formula for these operations is `= INT(2*RAND())`.

1. Enter the formula `= INT(2*RAND())` in cell A5 of a new workbook and copy this formula down to cell A1003 by selecting cell A4, then clicking the fill

| | A | B | C | D | E |
|---|---|---|---|---|---|
| 1 | *Simulation of 1000 Tosses of a Fair Coin* | | | | |
| 2 | | | | | |
| 3 | | 0 | | | |
| 4 | =INT(2*RAND()) | =A4+B3 | 1 | 0.5 | =B4/C4 |
| 5 | =INT(2*RAND()) | =A5+B4 | 2 | 0.5 | =B5/C5 |
| 6 | =INT(2*RAND()) | =A6+B5 | 3 | 0.5 | =B6/C6 |

Figure 4.1: Simulating 1000 Tosses of a Fair Coin

handle, and dragging to cell A1003 to generate 1000 tosses of a fair coin (0 representing tails and 1 representing heads).

2. Enter the value "0" in cell B3 followed by the formula "= A4+B3" in cell B4. Copy the formula in cell A4 down to cell A1003. Column B tracks the cumulative number of heads.

3. Enter the number 1 in cell C4 and fill to cell C1003 with successive integers $\{1, 2, \ldots, 1000\}$. This can be achieved efficiently by selecting cell C4 and then choosing **Edit – Fill – Series** from the Menu Bar. Complete the **Series** dialog box with Series in **Columns**, Type **Linear** and **Step Value** 1, **Stop Value** 1000. Click OK. Column C will label the 1000 tosses.

4. Fill cells D4 to D1003 with the value 0.5. This will represent the horizontal line at height 0.5 on the graph.

5. Enter the formula "=B4/C4" in cell E4 and copy to cell E1003.

Figure 4.1 shows part of the workbook with the required formulas. We next construct a graph displaying the same results.
Click the **ChartWizard** button.

1. **Users of Excel 5/95**

   - In Step 1 enter the data range C4:E1003.
   - In Step 2 click the **Line** chart type.
   - In Step 3 select Format **2**.
   - In Step 4 click the button for Data Series in **Columns**. Enter "1" for Use First 1 Column for Category(X) axis labels and enter "0" for Use First 0 Column for Legend Text.
   - In Step 5 click the radio button **No** for Add a legend?, and label the chart and X axis as shown in Figure 4.2. Click Finish.

   **Users of Excel 97/98**

   - In Step 1 click the **Line** chart type and the second Chart sub-type **Stacked Line**.

| | A | B | C | D | E |
|---|---|---|---|---|---|
| 1 | *Simulation of 1000 Tosses of a Fair Coin* | | | | |
| 2 | | | | | |
| 3 | | 0 | | | |
| 4 | 1 | 1 | 1 | 0.5 | 1.0000 |
| 5 | 1 | 2 | 2 | 0.5 | 1.0000 |
| 6 | 0 | 2 | 3 | 0.5 | 0.6667 |
| 7 | 0 | 2 | 4 | 0.5 | 0.5000 |
| 8 | 1 | 3 | 5 | 0.5 | 0.6000 |
| 9 | 0 | 3 | 6 | 0.5 | 0.5000 |
| 10 | 1 | 4 | 7 | 0.5 | 0.5714 |
| 11 | 0 | 4 | 8 | 0.5 | 0.5000 |
| 12 | 0 | 4 | 9 | 0.5 | 0.4444 |
| 13 | 0 | 4 | 10 | 0.5 | 0.4000 |
| 14 | 0 | 4 | 11 | 0.5 | 0.3636 |
| 15 | 1 | 5 | 12 | 0.5 | 0.4167 |
| 16 | 0 | 5 | 13 | 0.5 | 0.3846 |
| 17 | 1 | 6 | 14 | 0.5 | 0.4286 |
| 18 | 0 | 6 | 15 | 0.5 | 0.4000 |
| 19 | 1 | 7 | 16 | 0.5 | 0.4375 |
| 20 | 1 | 8 | 17 | 0.5 | 0.4706 |
| 21 | 0 | 8 | 18 | 0.5 | 0.4444 |
| 22 | 0 | 8 | 19 | 0.5 | 0.4211 |
| 23 | 1 | 9 | 20 | 0.5 | 0.4500 |
| 24 | 1 | 10 | 21 | 0.5 | 0.4762 |
| 25 | 0 | 10 | 22 | 0.5 | 0.4545 |
| 26 | 0 | 10 | 23 | 0.5 | 0.4348 |
| 27 | 1 | 11 | 24 | 0.5 | 0.4583 |

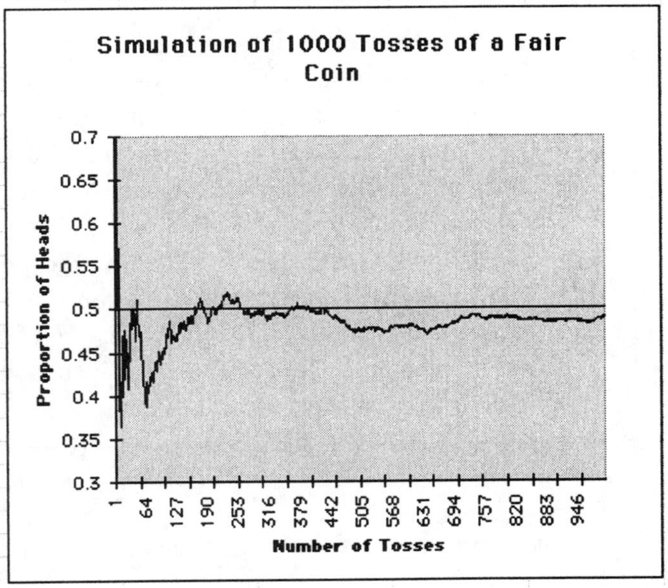

**Simulation of 1000 Tosses of a Fair Coin**

Proportion of Heads: 0.3, 0.35, 0.4, 0.45, 0.5, 0.55, 0.6, 0.65, 0.7

Number of Tosses: 1, 64, 127, 190, 253, 316, 379, 442, 505, 568, 631, 694, 757, 820, 883, 946

Figure 4.2: Law of Large Numbers

- In Step 2 on the **Data Range** tab enter D4:E1003 for the Data range check the radio button **Series in: Columns.**
- In Step 3 on the **Titles** tab, enter the title and labels of the axes, on the **Axes** tab check radio button **Automatic** for Category (X) axis and check the Value (Y) axis box, on the **Gridlines** tab turn off all gridlines, on the **Legend** tab clear the legend, and finally on the **Data Labels** tab select the radio button **None.**
- In Step 4 embed the graph in the current workbook. Click Finish.

2. Format the X and Y axes as shown in Figure 4.2, for instance by changing the number of categories between tick marks and reorienting the X-axis labels.

A segment of the completed workbook with the embedded graph is shown in Figure 4.2. Press the **F9** key to reevaluate all functions and the graph will dynamically change. From column A you can see the random sequence of heads and tails generated, while column E exhibits the proportions of heads. These are quite variable at first but then settle down, appearing to approach the value 0.5 (shown by the horizontal line). This behavior is known in statistics as a law of large numbers, commonly referred to as the "law of averages."

## Charting Shaq's Free Throws

**Example 4.2.** (Exercise 4.11 page 219 in the text.) The basketball player Shaquille O'Neal makes about half of his free throws over an entire season. Use Excel to simulate 100 free throws shot independently by a player who has probability 0.5 of making each shot. The technical term for independent trials with yes/no outcomes is Bernoulli trials. Our outcomes here are hit or miss.
(a) What percent of the 100 shots did he hit in the simulation?
(b) Examine the sequence of hits and misses. How long was the longest run of shots made? Of shots missed? (Sequences of random outcomes often show runs longer than our intuition thinks likely.)

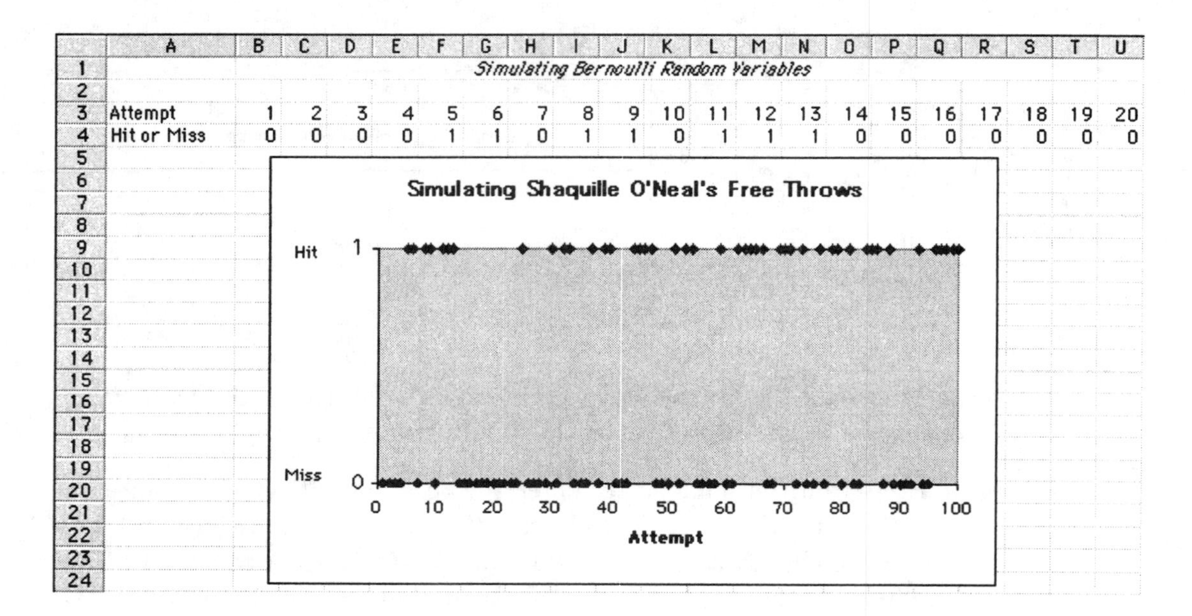


Figure 4.3: Simulating Shaq's Free Throws

**Solution.** Again we will use the `RAND()` function to generate a sequence of 100 free throws and then invoke the **ChartWizard** to dramatically display the results.

1. Enter the labels "Attempt" in Cell A3 of a new workbook and "Hit" or "Miss" in cell A4. Enter the number "1" in cell B3 and fill to cell CW3 with successive integers $\{1, 2, \ldots, 100\}$. This can be achieved efficiently by selecting cell B3, then choosing **Edit – Fill – Series...** from the Menu Bar. Complete the **Series** dialog box with **Series in** Rows, **Type** Linear, **Step value** 1, and **Stop value** 100. Click OK.

2. Enter the = `INT(2*RAND())` in cell B4. Select B4, click the fill handle, and drag to cell CW4 to generate 100 independent Bernoulli random variables.

In Figure 4.3 we show our workbook with the first 20 free throw simulations together with a graph of Shaq's hits or misses in 100 attempts. Notice how much more illuminating the graph is than the numerical sequence of hits and misses in showing random variation, including the presence of hot and cold streaks based on chance alone. The following steps explain how this chart is constructed.

Click the **ChartWizard** button.

1. **Users of Excel 5/95**

    - In Step 1 enter the data range A3:CW4.
    - In Step 2 click the **Scatter** chart type.
    - In Step 3 select Format **1**.
    - In Step 4 click the button for Data Series in **Rows**. Enter "1" for Use First Row for Category(X) Axis Labels and enter "1" for Use First 1 Column for Legend Text.
    - In Step 5 select the radio button **No** for Add a legend?, and label the chart and X axis as shown in Figure 4.3.
    - Click Finish.

    **Users of Excel 97/98**

    - In Step 1 click the **XY (Scatter)** chart type and the first Chart sub-type (upper left on right side).
    - In Step 2 on the **Data Range** tab enter A3:CW4 for the range and check the radio button **Rows** for Series in:.
    - In Step 3 on the **Titles** tab, enter the title and labels of the axes, on the **Axes** tab check both Category (X) axis, and Category (Y) axis, on the **Gridlines** tab turn off all gridlines, on the **Legend** tab clear the legend, and finally on the **Data Labels** tab select the radio button **None**.
    - In Step 4 embed the graph in the current workbook.
    - Click Finish.

2. After the chart appears embedded on your workbook select it for editing and add the text "Hit" and "Miss" on the vertical axis.

    (a) By entering = `SUM`(B4:CW4)/100 in an empty cell we find that the player hit 45% of his shots in the simulation.
    (b) From Figure 4.3 we read that the longest run of hits is 5 and the longest run of misses is 11.

To simulate an additional 100 free throws press the F9 key.

## 4.2 Probability Models

A probability model consists of a list of possible outcomes and a probability for each outcome (or interval of outcomes, in the case of continuous models). The probabilities are determined by the experiment which leads to the occurrence of one or more of the outcomes in the specified list.

Excel provides many distributions which may be constructed in a common fashion with the **Function Wizard** or the **Formula Palette**. The meaning of the required parameters is available online through Excel's help feature. Because of its prominence, the normal distribution was already discussed in Chapter 1. The binomial model will be considered in Chapter 5.

### Uniform

A uniform random number is one whose values are spread out uniformly across the interval from 0 to 1. Its density curve has height 1 over the interval 0 to 1.

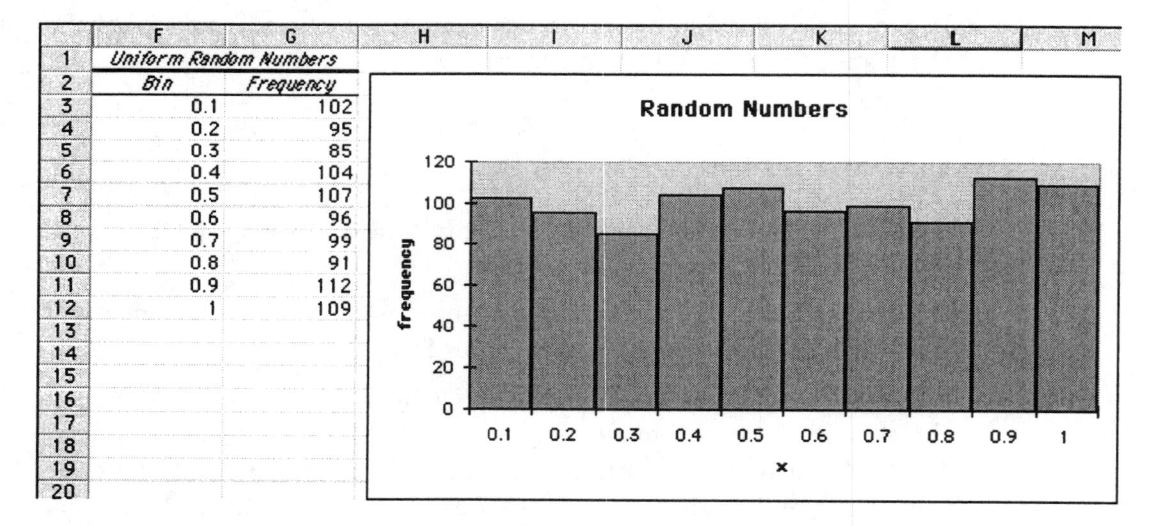


Figure 4.4: Simulating Uniform Random Variables

**Example 4.3.** (Exercise 4.22 page 229 in the text.) Let X be a uniform random number between 0 and 1. Use Excel to generate 1000 random uniform numbers and from your simulations estimate the following probabilities and compare them with the theoretical values.

(a) $P(0 \leq X \leq 0.4)$

(b) $P(0.4 \leq X \leq 1)$

(c) $P(0.3 \leq X \leq 0.5)$

(d) $P(0.3 < X < 0.5)$.

**Solution.** Use `RAND()` to generate 1000 uniform random variables in a column and construct a histogram with bin intervals of width 0.10 beginning at 0 and ending at 1. Figure 4.4 shows the sample output from a workbook where this has been done. The frequencies shown are the number of times the random number generator produced a number X in the specified interval. The values listed under the heading *Bin* are the right endpoints of the intervals. We count the number of observations in the relevant intervals and divide by 1000 to convert to a probability.

(a) $P(0 \leq X \leq 0.4) = 0.386$
(b) $P(0.4 \leq X \leq 1) = 0.614$
(c) $P(0.3 \leq X \leq 0.5) = 0.211$
(d) $P(0.3 < X < 0.5) = 0.211$.

The theoretical values are 0.4, 0.6, 0.2, and 0.2, respectively.

## Triangular – Adding Random Numbers

**Exercise 4.4.** (Exercise 4.23 page 229 in the text.) Generate two random numbers between 0 and 1 and take Y to be their sum. Clearly the sum Y can take any number between 0 and 2. It is known that the idealized density curve of Y is a triangle. Use Excel to generate 1000 pairs of uniform random numbers, add them, and from your simulations estimate the following probabilities and compare them with the theoretical values. See Figure 4.5.

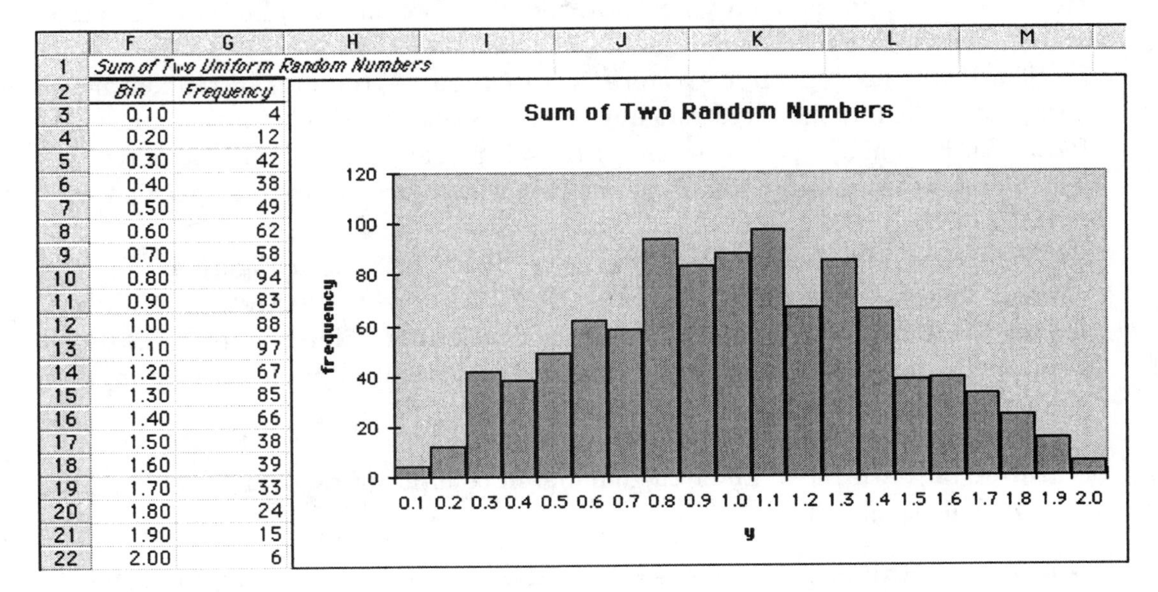

*Sum of Two Uniform Random Numbers*

| *Bin* | *Frequency* |
|---|---|
| 0.10 | 4 |
| 0.20 | 12 |
| 0.30 | 42 |
| 0.40 | 38 |
| 0.50 | 49 |
| 0.60 | 62 |
| 0.70 | 58 |
| 0.80 | 94 |
| 0.90 | 83 |
| 1.00 | 88 |
| 1.10 | 97 |
| 1.20 | 67 |
| 1.30 | 85 |
| 1.40 | 66 |
| 1.50 | 38 |
| 1.60 | 39 |
| 1.70 | 33 |
| 1.80 | 24 |
| 1.90 | 15 |
| 2.00 | 6 |



Figure 4.5: Simulating Triangular Random Variables

(a) $P(0 \leq X \leq 0.5)$
(b) $P(0.5 \leq X \leq 1.5)$.

**Solution.** Again use `RAND()` to generate 1000 pairs of uniform random variables in two columns, add the columns and construct a histogram with bin intervals of width 0.10 beginning at 0 and ending at 2. Figure 4.5 shows the sample output from a workbook where this has been done. The frequencies shown are the number of times the random number generator produced a number in the specified interval.

(a) $P(0 \leq X \leq 0.5) = 0.145$

(b) $P(0.5 \leq X \leq 1.5) = 0.738$.

The corresponding theoretical values are 0.125 and 0.750, respectively.

## 4.3 Sampling Distributions

We have seen how the Excel function `RAND()` picks a number uniformly on the interval (0, 1). To generate a uniform random variable on another interval $(a, b)$ use $= a+$`RAND`$*(b-a)$. Using the inverse probability function $h(a) = \inf\{x : F(x) \geq a\}$ we can then generate other distributions. Thus = `NORMINV(RAND()`, Mean, StDev) returns a random normal with mean given by Mean (either a numerical value or a named reference to a numerical value) and standard deviation StDev. By examining the list of functions available (clicking the button $f_x$ on the Standard Toolbar) you can determine which distributions Excel can simulate this way and how to describe the required parameters.

In addition to the function `RAND()`, Excel has a **Random Number Generation** tool built into the **Analysis ToolPak** which provides an alternate and more systematic approach to simulation.

The **Random Number Generation** tool creates columns of random numbers, as specified by the user, from any of six probability models (uniform, normal, Bernoulli, binomial, Poisson, discrete) as well as having an option for patterned which does not create random data but rather data according to a specified pattern.

All options are invoked from a common dialog box (as in Figure 4.7 for the normal) following the choice **Tools – Data Analysis – Random Number Generation** from the Menu Bar. Select the distribution of interest using the drop-down arrow and the Parameters sub-box will automatically change, prompting input of parameters.

**Number of Variables.** Enter the number of columns of random variables. The default is all columns.

**Number of Random Numbers.** Enter the number of rows (cases) of random variables.

**Distributions.** Use the drop down arrow to open a list of choices with requested parameters.

**Uniform.** upper and lower limits

**Normal.** $\mu, \sigma$

**Bernoulli.** $p =$ probability of success which Excel unfortunately refers to this as a p Value.

**Binomial.** $p, n$

**Poisson.** $\lambda$

**Discrete.** Specify the possible values and their corresponding probabilities. Before using this option enter the values and probabilities in adjacent columns in the workbook.

**Patterned.** This option creates data according to a prescribed pattern of values repeated in specified steps. This is useful if a linear array of data needs to be coded using another variable.

## Constructing a Sampling Distribution

**Example 4.4.** (Example 4.12 page 239 in text.) Sulfur compounds such as dimethyl sulfide (DMS) are sometimes present in wine. DMS causes "offodors" in wine, so winemakers want to know the odor threshold, the lowest concentration of DMS that the human nose can detect. Different people have different thresholds and extensive studies have found that the DMS odor threshold of adults follows roughly a normal distribution with mean $\mu = 25$ micrograms per liter and standard deviation $\sigma = 7$ micrograms per liter. Take 1000 samples of size 10. For each sample find $\bar{x}$ the sample mean and then make a histogram of these 1000 values.

| | A | B | C | D | E | F | G | H | I | J | K |
|---|---|---|---|---|---|---|---|---|---|---|---|
| 1 | | | | | *Constructing a Sampling Distribution* | | | | | | |
| 2 | | | | | | | | | | | *Means* |
| 3 | 12.845 | 19.261 | 30.999 | 17.543 | 21.822 | 23.399 | 24.207 | 26.528 | 20.657 | 21.226 | 21.849 |
| 4 | 26.788 | 24.658 | 23.261 | 25.003 | 23.272 | 24.115 | 31.473 | 26.214 | 28.371 | 17.199 | 25.035 |
| 5 | 33.669 | 28.681 | 28.678 | 20.911 | 26.808 | 26.646 | 24.334 | 25.482 | 21.488 | 21.047 | 25.774 |
| 6 | 22.058 | 21.025 | 27.947 | 26.715 | 15.537 | 24.991 | 28.966 | 19.112 | 26.708 | 25.152 | 23.821 |
| 7 | 27.996 | 25.786 | 24.842 | 14.356 | 25.984 | 26.865 | 18.778 | 20.431 | 13.106 | 28.323 | 22.647 |
| 8 | 18.091 | 25.692 | 24.541 | 11.894 | 26.164 | 21.968 | 19.545 | 23.493 | 29.451 | 33.636 | 23.448 |
| 9 | 43.398 | 14.449 | 16.291 | 21.091 | 24.120 | 18.500 | 19.726 | 29.149 | 31.402 | 22.831 | 24.096 |
| 10 | 30.094 | 19.657 | 28.159 | 24.365 | 33.452 | 18.317 | 25.684 | 28.610 | 20.803 | 20.534 | 24.968 |
| 11 | 29.763 | 23.559 | 10.042 | 22.239 | 27.822 | 23.471 | 25.232 | 28.800 | 27.057 | 30.221 | 24.821 |
| 12 | 14.968 | 27.973 | 30.785 | 31.336 | 32.244 | 24.494 | 27.042 | 24.557 | 11.878 | 8.158 | 23.343 |

Figure 4.6: Samples of size 10 from $N(25, 7)$ with Sample Means

**Solution.**

1. Choose **Tools – Data Analysis – Random Number Generation** from the Menu Bar and complete the dialog box as in Figure 4.7. The output will appear in cells A3:J1002 (ten columns with 1000 rows in each). Each row represents a sample of size 10.

Random Number Generation

Input
Number of variables: 10
Number of random numbers: 1000
Random seed:
Distribution: Normal
Mean = 25
Standard deviation = 7

OK
Cancel
Help

Output Options
Output range: $A$3
New worksheet ply:
New workbook

Figure 4.7: Random Number Generation Tool – Normal

2. Enter the formula = `AVERAGE`(A3:J3) in cell K3 to compute the average of the first sample of size 10.

3. Select cell K3, click the fill handle, and drag down to cell K1002 to obtain the averages of all the rows. See Figure 4.6 showing the first 10 samples. Since these numbers are random your output will of course be different.

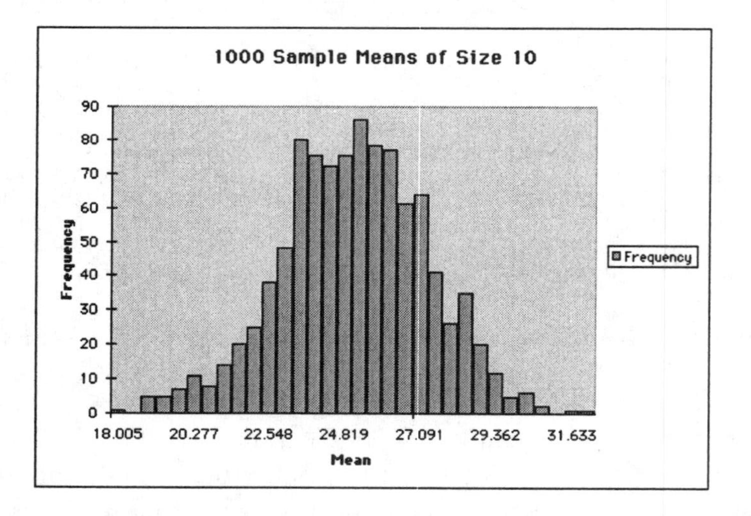


Figure 4.8: Histogram – 1000 Samples of Size 10

4. From the Menu Bar choose **Tools** – **Data Analysis** and scroll to **Histogram.** Construct a histogram of the data in column K. See Figure 4.8.

# Chapter 5

# Probability Theory

In Chapter 4 we used simulation as an approach to understanding variation. Probability models are theoretical descriptions of variation. An important class of models arises when the data are counts of some variable. This leads to the binomial model for sample counts and sample proportions. Its distribution can be approximated by normal curves and directly demonstrates several important results about sample means $\bar{x}$ in general:

1. $\bar{x}$ is an unbiased estimate of the population mean $\mu$;
2. The standard deviation of $\bar{x}$ is equal to $\frac{\sigma}{\sqrt{n}}$ where $n$ is the sample size and $\sigma$ the population standard deviation;
3. The sampling distribution of $\bar{x}$ is approximately $N(\mu, \frac{\sigma}{\sqrt{n}})$.

## 5.1 General Probability Rules

Many probability models are derived from independent outcomes. The multiplication rule says that the probability of two independent events occurring is the product of their individual probabilities.

> **Example 5.1.** (Based on Exercise 5.17 page 269 in text.)
>
> (a) What is the probability that a pair of fair dice will show the same number face up when they are rolled?
>
> (b) Simulate rolling a pair of fair dice 500 times and compare your empirical results with the theoretical values.

**Solution.**
(a) The solution requires use of both the addition rule and the multiplication rule. The chance that both dice show the same *specified* value, say a one-spot, face up is $\frac{1}{6} \times \frac{1}{6} = \frac{1}{36}$. But there are six possible values for the number shown. Hence we must add $\frac{1}{36}$ six times to get $\frac{1}{6}$.

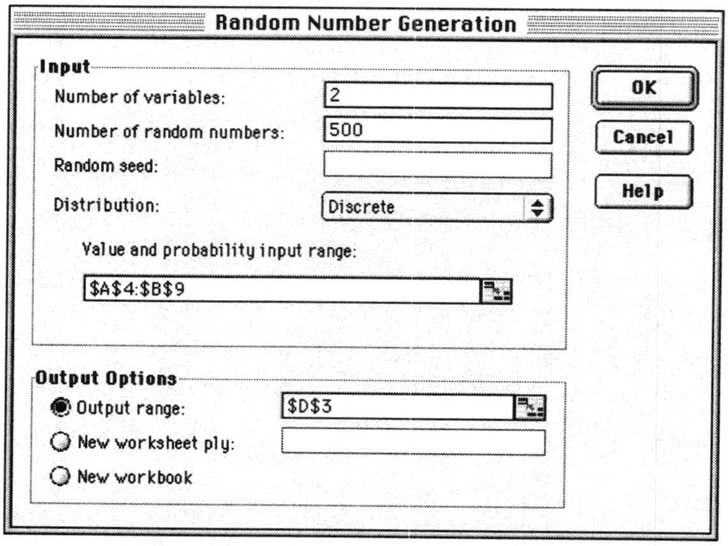

**Random Number Generation**

**Input**
Number of variables: 2
Number of random numbers: 500
Random seed:
Distribution: Discrete
Value and probability input range: $A$4:$B$9

**Output Options**
(•) Output range: $D$3
( ) New worksheet ply:
( ) New workbook

OK | Cancel | Help

Figure 5.1: Random Number Generation Tool

(b) Enter $\{1, 2, \ldots, 6\}$ into cells A4:A9 and enter $\{1/6, 1/6, \ldots, 1/6\}$ into cells B4:B9 of a workbook. (Note: Excel may interpret the value 1/6 as a date 6-Jan. If this happens then format the cells by choosing **Format** – **Cells** from the Menu Bar and selecting the Number tab.) Then choose **Tools** – **Data Analysis** – **Random Number Generation** from the Menu Bar and complete as in Figure 5.1. The output will appear in cells D3:E502 (Figure 5.2). Next use the conditional `IF` function to count when both dice showed the same value. Do this by inserting the formula = `IF`(D3=E3,1,0) in cell F3 and filling the formula down to cell E502. Finally, place the formula = `SUM`(F3:F502)/500 in a free cell to find the proportion out of the 500 trials when the two dice showed the same number face up. The simulation gives 0.15, while the theoretical value is $\frac{1}{6} = 0.17$.

| | A | B | C | D | E | F | G | H | I |
|---|---|---|---|---|---|---|---|---|---|
| 1 | *Simulation of a Pair of Fair Dice* | | | | | | | | |
| 2 | | | | Die 1 | Die2 | | | | Enter =IF(D3=E3, 1,0) in cell F3 |
| 3 | k | P(X=k) | | 3 | 1 | 0 | | | then fill to cell F502. |
| 4 | 1 | 0.16667 | | 1 | 4 | 0 | | | |
| 5 | 2 | 0.16667 | | 5 | 5 | 1 | | | Enter =SUM(F3:F502)/500 in cell I7 |
| 6 | 3 | 0.16667 | | 2 | 5 | 0 | | | to get p. |
| 7 | 4 | 0.16667 | | 5 | 3 | 0 | | p= | 0.152 |
| 8 | 5 | 0.16667 | | 4 | 5 | 0 | | | |
| 9 | 6 | 0.16667 | | 2 | 4 | 0 | | | |
| 10 | | | | 1 | 2 | 0 | | | |
| 11 | | | | 6 | 1 | 0 | | | |
| 12 | | | | 2 | 6 | 0 | | | |

Figure 5.2: Simulating a Pair of Fair Dice

## 5.2 The Binomial Distributions

A binomial distribution is associated with an experiment comprised of $n$ independent trials each of which has the same success probability $p$. The random variable $X$ counts the number of successes.

It is known that

$$P(X = k) = \binom{n}{k} p^k (1-p)^{n-k} \qquad k = 0, 1, 2, \ldots, n$$

$$\text{mean} = \mu_X = np$$

and

$$\text{standard deviation} = \sigma_X = \sqrt{np(1-p)} .$$

The corresponding Excel function is BINOMDIST($k, n, p$, cumulative). If the parameter cumulative is set to "false," Excel returns the probabilities $P(X = k)$, while if it is set to "true" Excel returns the cumulative probabilities $P(X \leq k)$.

**Example 5.2.** Construct a binomial table for $n = 15$ and $p = 0.03$, including both individual and cumulative probabilities.

**Solution.**

1. Enter the label $k$ in cell A1 and the label $P(X = k)$ in cell B1 of a new workbook. In A2:A17 enter the values $\{0, 1, 2, \ldots, 15\}$.

2. Activate cell B2. Using either the **Function Wizard** or the **Formula Palette** construct the binomial function by selecting **Statistical** for Function Category, and BINOMDIST for Function Name.

3. Input the following into the dialog box.

   number_s. Enter the cell address A2.
   trials. Enter the value 15.
   probability_s. Enter the value 0.3.
   cumulative. Enter the value 0.

   Click Finish or OK.

4. Activate cell B2, click the fill handle in the lower right corner and drag to cell B17 to fill the column with individual binomial probabilities (Figure 5.3).

5. Next label cell C1 as $P(X <= k)$ and repeat Steps 2, 3, and 4. Activate C2 instead of B2 in Steps 2 and 4 and enter the value 1 for the cumulative distribution in Step 3.

The resulting table of individual and cumulative binomial probabilities appears in Figure 5.3.

| | A | B | C |
|---|---|---|---|
| 1 | x | P(X=k) | P(X<=k) |
| 2 | 0 | 0.00475 | 0.00475 |
| 3 | 1 | 0.03052 | 0.03527 |
| 4 | 2 | 0.09156 | 0.12683 |
| 5 | 3 | 0.17004 | 0.29687 |
| 6 | 4 | 0.21862 | 0.51549 |
| 7 | 5 | 0.20613 | 0.72162 |
| 8 | 6 | 0.14724 | 0.86886 |
| 9 | 7 | 0.08113 | 0.94999 |
| 10 | 8 | 0.03477 | 0.98476 |
| 11 | 9 | 0.01159 | 0.99635 |
| 12 | 10 | 0.00298 | 0.99933 |
| 13 | 11 | 0.00058 | 0.99991 |
| 14 | 12 | 0.00008 | 0.99999 |
| 15 | 13 | 0.00001 | 1.00000 |
| 16 | 14 | 0.00000 | 1.00000 |
| 17 | 15 | 0.00000 | 1.00000 |

Figure 5.3: Binomial Probabilities

## Binomial Distribution Histogram

We can quickly construct a histogram using the **ChartWizard** displaying the binomial probabilities just calculated. As the procedure is identical to earlier constructions of charts we omit the details. This histogram appears in Figure 5.4.

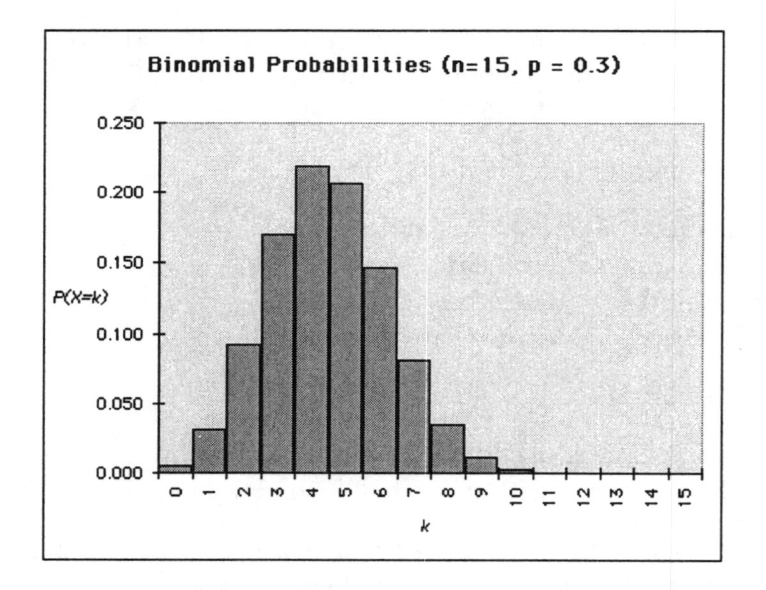


Figure 5.4: Binomial Histogram

### Inverse Cumulative Binomial

CRITBINOM(trials, probability_s, alpha) returns the smallest $x$ for which the binomial cumulative distribution function (c.d.f.) is greater than or equal to alpha, that is if $B(x)$ represents the binomial c.d.f. then = CRITBINOM($n, p, \alpha$) returns

$$B^{-1}(\alpha) = \inf\{x : B(x) \geq \alpha\}, \quad 0 < \alpha \leq 1.$$

For $\alpha = 0$ this definition gives $-\infty$ and Excel gives the error message #NUM!.

Inverse probabilities are useful for finding $P$-values and in **simulation** because from the definition, if $U$ is uniform $(0, 1)$ and $F(x)$ is an arbitrary c.d.f. with inverse defined by

$$F^{-1}(\alpha) = \inf\{x : F(x) \geq \alpha\}$$

then

$$X = F^{-1}(U)$$

has the specified distribution $F(x)$. Thus, for instance

$$= \texttt{CRITBINOM}(u, p, U)$$

is a binomial random variable on $n$ trials and success probability $p$, and

$$= \texttt{NORMINV}(U, \mu, \sigma)$$

is a $N(\mu, \sigma)$ random variable.

## Normal Approximation to Binomial

The formula for binomial probabilities becomes awkward as the number of trials $n$ increases. But there is an approximation based on the normal distribution: ***As the number of trials n gets larger, the binomial distribution gets close to a normal distribution.*** This is known as a central limit theorem.

**Example 5.3.** A roulette wheel has 38 slots – 8 are black, 18 are red, and 2 are green. When the wheel is spun, a ball is equally likely to come to rest in any of the slots. Gamblers can place a number of different bets in roulette. One of the simplest wagers chooses red or black. A bet of one dollar on red will pay off an additional dollar if the ball lands in a red slot. Otherwise the player loses his dollar. When a gambler bets on red or black, the two green slots belong to the house. A gambler's winnings on a \$1 bet are either \$1 or −\$1.

(a) Simulate a gambler's winnings after 50 bets and compare the gambler's mean winnings per bet with the theoretical results.

(b) Compare the results with the normal approximation.

**Solution.** The number of wins after 50 bets $X$ is a binomial $B(50, 10/38)$ random variable with

$$\text{mean} \qquad \mu_X = 50\left(\frac{18}{38}\right) = 23.684$$

$$\text{standard deviation} \qquad \sigma_X = \sqrt{50\left(\frac{18}{38}\right)\left(\frac{20}{38}\right)} = 3.5306$$

The proportion of wins after 50 bets is $\hat{p} = X/50$ with

$$\text{mean} \qquad \mu_{\hat{p}} = \frac{18}{38} = 0.4737$$

$$\text{standard deviation} \qquad \sigma_{\hat{p}} = \sqrt{\left(\frac{18}{38}\right)\left(\frac{20}{38}\right)\Big/50} = 0.0706$$

Since the gambler wins \$1 or loses \$1 his average winnings per game, denoted by $\bar{w}$, is $\bar{w} = \hat{p}(1) + (1 - \hat{p})(-1) = 2\hat{p} - 1$ with

$$\text{mean} \qquad \mu_{\bar{w}} = 2\mu_{\hat{p}} - 1 = -0.0527$$

$$\text{standard deviation} \qquad \sigma_{\bar{w}} = 2\sigma_{\hat{p}} = 0.14123$$

By the **normal approximation to the binomial** $\hat{p}$ is approximately normal, and therefore so is $\bar{w}$ and

$$\bar{w} \text{ is approximately } N(-0.0527, 0.14123)$$

We can simulate a binomial random variable $X$, convert it first to $\hat{p} = \frac{X}{n}$ and then to $\bar{w} = 2\hat{p} - 1$ after which we construct a histogram of the simulation results.

The following steps, referring to Figure 5.5, show how to develop a workbook to simulate 500 replications of 50 games. We will use the `RAND` function which *links* the output to a histogram. By using the F9 key you can repeat the simulation and watch how the results vary.

We will use bin intervals determined by the simulation and boundaries located at multiples of the standard deviation from the mean in order to allow comparison with the 68-95-99.7% rule mentioned in Chapter 1. We use the sample mean $\bar{w}$ and sample standard deviation $s_{\bar{w}}$ rather than the theoretical values $\mu_{\bar{w}}$ and $\sigma_{\bar{w}}$ respectively, in order to show how $\bar{w}$ and $s_{\bar{w}}$ vary each time the F9 key is pressed.

1. Prepare a new workbook by entering "Demonstrating the Normal Approximation to the Binomial" in cell A1 and centering the heading across A1:E1. Enter "Simulation Number" in A3:A4 and "Average per Game" in B3:B4. Enter "mean" in C2, "st_dev" in C3, select C2:D3, and from the Menu Bar choose **Insert** – **Name** – **Create** and check the box **Left Column** in the dialog box. The formulas required are

   mean:= `AVERAGE`(B5:B1005)

   and

   st_dev:= `SQRT`((1 – mean*mean)/50)

| | A | B | C | D | E | F | G |
|---|---|---|---|---|---|---|---|
| 1 | | | *Demonstrating the Normal Approximation to the Binomial* | | | | |
| 2 | | | mean | -0.0526 | true mean = | -0.0527 | |
| 3 | *Simulation* | *Average* | st_dev | 0.14123 | true st_dev = | 0.14123 | |
| 4 | *Number* | *Per Game* | *Formula entered in column B* | =2*CRITBINOM(50,18/38,RAND())/50 -1 | | | |
| 5 | 1 | 0.160 | *Bin Formulas* | *Bin* | *Freq.* | | |
| 6 | 2 | 0.040 | =mean-3.5*st_dev | -0.55 | 0 | | |
| 7 | 3 | -0.080 | =mean-3*st_dev | -0.48 | 2 | | |
| 8 | 4 | -0.080 | =mean-2.5*st_dev | -0.41 | 1 | | |
| 9 | 5 | 0.040 | =mean-2*st_dev | -0.34 | 10 | | |
| 10 | 6 | 0.040 | =mean-1.5*st_dev | -0.26 | 19 | | |
| 11 | 7 | 0.080 | =mean-st_dev | -0.19 | 55 | | |
| 12 | 8 | -0.080 | =mean-0.5*st_dev | -0.12 | 46 | | |
| 13 | 9 | -0.120 | =mean | -0.05 | 113 | | |
| 14 | 10 | -0.160 | =mean+0.5*st_dev | 0.02 | 104 | | |
| 15 | 11 | -0.120 | =mean+st_dev | 0.09 | 81 | | |
| 16 | 12 | 0.000 | =mean+1.5*st_dev | 0.16 | 23 | | |
| 17 | 13 | -0.240 | =mean+2*st_dev | 0.23 | 29 | | |
| 18 | 14 | 0.240 | =mean+2.5*st_dev | 0.30 | 15 | | |
| 19 | 15 | -0.160 | =mean+3*st_dev | 0.37 | 2 | | |
| 20 | 16 | 0.200 | =mean+3.5*st_dev | 0.44 | 0 | | |

Figure 5.5: Simulating the Normal Approximation to the Binomial

2. Enter the labels "Bin" in D5 and "Freq." in E5.

3. Next we describe the bin endpoints in cells D6:D19 based on the sample mean and sample standard deviation. Enter "= mean −3.5∗ st_dev" in D6, " = mean −3.0∗ st_dev" in D7 and so on. Refer to Figure 5.5 where we have shown in cells C6:C19 the formulas to be entered in D6:D19. Note that you do not require a column C.

4. Enter the values 1, 2, . . . , 500 in cells A5:A504 as follows: Enter "1" in A5. Select A5 and choose **Edit – Fill – Series...** from the Menu Bar. In the **Series** dialog box check Series in **Columns** and Type **Linear**. Clear the **Trend** box and type "1" and "600" for the **Step** and **Stop** values, respectively.

5. In cell B5 enter = 2*`CRITBINOM`(50,18/38, `RAND()`)/(50 − 1) to generate the random variable $\bar{w}$. We have shown the formula on line 4 beginning in column D. Select B5, click the fill handle at the lower right corner of B5 and drag down to cell B504. Cells B5:B504 are now filled with 500 replications of the gambler's average net gain per game after 50 games.

6. Select E6:E19. Then type = `FREQUENCY`(B5:B504,D6:D19) in the entry area of the **Formula Bar**. Hold down the **Shift and Control** keys (either **Macintosh or Windows**) and press enter/return to **array-enter** the formula. The formula will appear **surrounded by braces { }** in the **Formula Bar** and the bin frequencies will appear in cells E6:E19.

7. Select cells D6:D19 and then complete the sequence of steps in the **ChartWizard** as discussed previously. The resulting histogram appears in Figure 5.6.

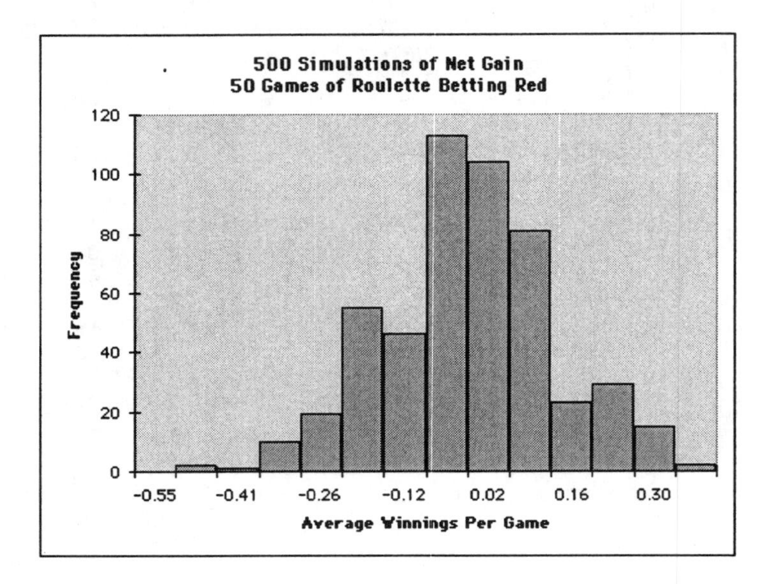


Figure 5.6: Graphing Outcomes of 50 Roulette Games

## Excel Output

The sample mean and sample standard deviation appear in D2:D3 and the population mean and standard deviation appear in F2:F3 for comparison purposes. For the simulation shown

$$\bar{w} = -0.0536 \qquad \mu_{\bar{w}} = -0.0527$$
$$s_{\bar{w}} = 0.14122 \qquad \sigma_{\bar{w}} = 0.14123$$

The table of frequency counts appears in E5:E19 with corresponding histogram in Figure 5.6. The histogram appears normal shaped with no unusual features.

Recalling that the bin entries are the right endpoints of the bin interval, we can determine the proportion of counts within 1, 2, and 3 standard deviation units of the mean. The simulation results are alarmingly good.

| | Actual | Theoretical |
|---|---|---|
| Within 1 s | .642 | .68 |
| Within 2 s | .944 | .95 |
| Within 3 s | .996 | .997 |

# Chapter 6

# Introduction to Inference

Statistical inference provides methods for drawing conclusions about a population from sample data. A sample provides the responses or measurements of individuals within the sample. A different sample will yield different data, and possibly suggest different conclusions about the populations.

This chapter introduces the two most common statistical procedures, confidence intervals and tests of significance for the mean of a population. These procedures are based on the normal distribution and hence are applicable only when the underlying population may be assumed to be approximately normal or when the sample size is so large that the central limit theorem justifies use of the normal distribution.

The main idea is the recognition that estimates based on the data have a sampling distribution which determines their variability in repeated sampling. In the setting of the normal distribution, this variability is completely captured by the standard deviation or estimated standard deviation.

## 6.1 Estimating with Confidence

A level $C$ confidence interval for a population mean $\mu$ is given by

$$\bar{x} \pm z^* \frac{\sigma}{\sqrt{n}}$$

where $\bar{x} = \frac{1}{n}\sum_{i=1}^{n} x_i$ is the sample mean, $n$ is the sample size, and $z^*$ is the value such that the area between $-z^*$ and $z^*$ under a standard normal curve equals $C$. The data $\{x_1, \ldots, x_n\}$ are assumed to come from an $N(\mu, \sigma)$ population with mean $\mu$ and a known standard deviation $\sigma$.

The Excel function required is `NORMSINV(`$a$`)` which returns the inverse of the standard normal cumulative distribution function $\Phi^{-1}(a)$. Thus for level $C$ confidence we use

$$z^* = \texttt{NORMSINV}\left(0.50 + \frac{C}{2}\right)$$

| | A | B | C | D |
|---|---|---|---|---|
| 1 | | *Confidence Interval for a Normal Mean* | | |
| 2 | | | | |
| 3 | | values | | formulas |
| 4 | User Input | | | |
| 5 | sigma | 0.0068 | | 0.0068 |
| 6 | conf | 0.99 | | 0.99 |
| 7 | Summary Statistics | | | |
| 8 | n | 3 | | =COUNT(Data) |
| 9 | xbar | 0.8404 | | =AVERAGE(Data) |
| 10 | Calculations | | | |
| 11 | SE | 0.0039 | | =sigma/SQRT(n) |
| 12 | z | 2.576 | | =NORMSINV(0.5+conf/2) |
| 13 | ME | 0.0101 | | =z*SE |
| 14 | Excel ME | 0.0101 | | =CONFIDENCE(1-conf,sigma,n) |
| 15 | Confidence Limits | | | |
| 16 | lower | 0.8303 | | =xbar-ME |
| 17 | upper | 0.8505 | | =xbar+ME |
| 18 | | | | |
| 19 | Data | | | |
| 20 | 0.8403 | | | |
| 21 | 0.8363 | | | |
| 22 | 0.8447 | | | |

Figure 6.1: Confidence Interval for a Normal Mean

**Example 6.1.** (Example 6.4 page 306 in text.) A manufacturer of pharmaceutical products analyzes a specimen from each batch of a product to verify the concentration of the active ingredient. The chemical analysis is not perfectly precise. Repeated measurements on the same specimen give slightly different results but the results of repeated measurements follow a normal distribution quite closely. The analysis procedure has no bias, so the mean $\mu$ of the population of all measurements is the true concentration in the specimen. The standard deviation of this distribution is known to be $\sigma = 0.0068$ grams per liter. The laboratory analyzes each specimen three times and reports the mean result.

Three analyses of one specimen give concentrations

$$0.8403 \qquad 0.8383 \qquad 0.8447$$

Give a 95% confidence interval for the true concentration $\mu$.

**Solution.** In Figure 6.1 the Excel formulas required are given in column D. These are entered into the adjacent cells in column B to create the workbook template to solve this problem. The Excel output is in column B. The user inputs required are the standard deviation and the confidence level. If the data have already been summarized then you can enter the values for the sample size $n$ and the average in cells B8:B9, respectively. Otherwise Excel will read the data and calculate $n$ and $\bar{x}$. The data can be located in a convenient place on the same sheet or it can be located on another sheet. The latter is particularly useful for large data sets. In this example with only three data points we have recorded them on the same sheet as the calculations.

The following steps describe how to construct the workbook.

1. Enter the labels as shown in column A.

2. **Name** the cell ranges to be used. Select cells A5:B6, A8:B9, A11:B13 and A16:B17. To select **non-contiguous blocks** of cells, make the first selection A5:B6, then hold down the **Control (Windows)** or **Command (Macintosh)** keys while selecting the other ranges. From the Menu Bar choose **Name – Create**, select **Left Column**, and then click OK. Next, select A19:A22 and from the Menu Mar choose **Name – Create**, select **Top Row**, and then click OK to name the data range.

3. Enter the above formulas into columns B8:B9, B11:B14 and B16:B17 and enter the data in A20:A22. Since you have named the data range you can refer to the cells A20:A22 as "Data", for instance as in the formula = `COUNT`(Data). Otherwise, you would type = `COUNT`(A20:A22) giving the actual locations. These formulas are sufficiently simple that you can enter them by hand rather than use the Function Wizard or Formula Palette.

4. Finally, for presentation purposes you can format the workbook as indicated.

The only input needed once the workbook has been constructed are the population standard deviation (sigma) and the confidence level. Type "0.0068" and "0.99" into cells B5 and B6, respectively. The results are immediately recorded in cells B16:B17 showing a lower confidence limit 0.8303 and an upper confidence limit 0.8505 for the population mean $\mu$.

## Explanation

The formula = `COUNT`(Data) gives the sample size by counting the number of cells named by the variable Data. You could also type the integer "3" instead. Likewise, = `AVERAGE`(Data) is the Excel formula for the sample mean $\bar{x}$. We have also included, for comparison purposes, the Excel formula = `CONFIDENCE`($\alpha, \sigma, n$), which calculates the margin of error (half width of interval) associated with a $C = 1 - \alpha$ level confidence interval for the mean of a normal distribution with standard deviation $\sigma$ based on a sample of size $n$. This function can be used in a less formal setting.

## The Meaning of a Confidence Interval

A confidence interval is a random interval which has a specified probability of containing an unknown parameter. Thus, a 90% confidence interval for a population mean has probability 0.90 of containing the mean. So, in repeated confidence intervals, in the long run approximately 90% of these confidence intervals would contain the population mean.

> **Example 6.2.** Take 100 SRS of size 3 from an $N(3.0, 0.2)$ population and construct a 90% confidence interval for the mean. Count how many times the confidence interval contains the mean 3.0.

**Solution.**

1. Following the instructions given in Example 4.4 for simulating samples from a specified distribution choose **Tools** – **Data Analysis** – **Random Number Generation** from the Menu Bar, complete a box like the one shown in Figure 4.7, with "3" for the **Number of Variables**, "100" for the **Number of Random Numbers**, "3.0" for the **Mean**, "0.2" for the **Standard Deviation**, and choose a convenient range for the output. In Figure 6.1 we have selected the range A8:C107.

2. In cell E8 enter
   = AVERAGE(A8:C8) – NORMSINV(0.5+0.9/2)*0.2/SQRT(3)
   In cell F8 enter
   = AVERAGE(A8:C8) + NORMSINV(0.5+0.9/2)*0.2/SQRT(3)

3. Select cells E8:F8, click the fill handle and drag the contents to F107. The cells in column F will contain the value 1 if the confidence interval for the data in the corresponding row contains the true value 3.0, while the cells will contain 0 otherwise.

4. Count the number of times 1 appears by entering = SUM(G8:G107) in an empty cell (H8, for example).

| | A | B | C | D | E | F | G | H |
|---|---|---|---|---|---|---|---|---|
| 1 | *Behavior of Repeated Confidence Intervals* | | | | | | | |
| 2 | | | | | | | | |
| 3 | lower limit=AVERAGE(A8:C8) -NORMSINV(0.5+0.90/2)*0.2/SQRT(3) | | | | | | | |
| 4 | upper limit=AVERAGE(A8:C8) +NORMSINV(0.5+0.90/2)*0.2/SQRT(3) | | | | | | | |
| 5 | G8 contains =IF(AND(E8<3, 3<F8), 1,0) | | | | | | | |
| 6 | | | | | lower | upper | | |
| 7 | | | | | | | | |
| 8 | 3.1772 | 2.7218 | 3.3097 | | 2.880 | 3.259 | 1 | 92 |
| 9 | 3.0863 | 3.0417 | 2.8220 | | 2.793 | 3.173 | 1 | |
| 10 | 2.8207 | 2.8480 | 2.9353 | | 2.678 | 3.058 | 1 | |
| 11 | 2.9131 | 3.2380 | 3.0292 | | 2.870 | 3.250 | 1 | |
| 12 | 3.0904 | 3.1497 | 2.9295 | | 2.867 | 3.246 | 1 | |
| 13 | 2.8767 | 3.0868 | 3.2555 | | 2.883 | 3.263 | 1 | |
| 14 | 2.8937 | 2.7254 | 2.9995 | | 2.683 | 3.063 | 1 | |
| 15 | 3.1976 | 3.0303 | 2.8750 | | 2.844 | 3.224 | 1 | |
| 16 | 2.8378 | 2.9206 | 2.7565 | | 2.648 | 3.028 | 1 | |
| 17 | 2.7972 | 2.9133 | 3.1956 | | 2.779 | 3.159 | 1 | |

Figure 6.2: Repeated Confidence Intervals

Figure 6.2 shows a portion of a workbook with the simulation for which 92 times out of 100 the true mean was within the 90% confidence limits.

| | A | B | C |
|---|---|---|---|
| 1 | | *Z Test for a Normal Mean* | |
| 2 | | | |
| 3 | | | |
| 4 | User Input | | |
| 5 | sigma | | Data |
| 6 | null | | 299.4 |
| 7 | alpha | | 297.7 |
| 8 | alternate | | 301.0 |
| 9 | Summary Statistics | | 298.9 |
| 10 | n | =COUNT(Data) | 300.2 |
| 11 | xbar | =AVERAGE(Data) | 297.0 |
| 12 | Calculations | | |
| 13 | SE | =sigma/SQRT(n) | |
| 14 | z | =(xbar-Null)/SE | |
| 15 | Lower Test | | |
| 16 | lower_z | =NORMSINV(alpha) | |
| 17 | Decision | =IF(z<lower_z,"Reject H0","Do Not Reject H0") | |
| 18 | Pvalue | =NORMSDIST(z) | |
| 19 | Upper Test | | |
| 20 | upper_z | =-NORMSINV(alpha) | |
| 21 | Decision | =IF(z>upper_z,"Reject H0","Do Not Reject H0") | |
| 22 | Pvalue | =1-NORMSDIST(z) | |
| 23 | Two-Sided Test | | |
| 24 | two_z | =ABS(NORMSINV(alpha/2)) | |
| 25 | Decision | =IF(ABS(z)>two_z,"Reject H0","Do Not Reject H0") | |
| 26 | Pvalue | =2*(1-NORMSDIST(ABS(z))) | |

Figure 6.3: Significance Test for a Normal Mean – Formulas

## 6.2 Tests of Significance

Significance tests are used to judge whether a specified (null) hypothesis is consistent with a data set.

We create a workbook for testing the null hypothesis $H_0 : \mu = \mu_0$ for a specified null value $\mu_0$, against one-sided or two-sided alternatives. The data $\{x_1, x_2, \ldots, x_n\}$ are assumed to come from an $N(\mu, \sigma)$ population where $\sigma$ is known. The same procedure can also be used to carry out a large sample test. In the workbook below, the user can either test at a specified level of significance or determine a $P$-value.

The user inputs are the sample size, sample mean, standard deviation (which may be input as values, as formulas, or as named references depending on the context), null hypothesis, and level of significance.

**Example 6.3.** (Exercise 6.36 page 333 in text.) Bottles of a popular cola drink are supposed to contain 300 ml of cola. There is some variation from bottle to bottle because the filling machinery is not perfectly precise. The distribution of the contents is normal with standard deviation $\sigma = 3$ ml. A student who suspects that the bottle is underfilling measures the contents of six bottles. The results are

299.4 297.7 310.0 298.9 300.2 297.0.

| | A | B | C |
|---|---|---|---|
| 1 | *Z Test for a Normal Mean* | | |
| 2 | | | |
| 3 | | | |
| 4 | User Input | | |
| 5 | sigma | 3.0 | Data |
| 6 | null | 300.0 | 299.4 |
| 7 | alpha | 0.05 | 297.7 |
| 8 | alternate | lower | 301.0 |
| 9 | Summary Statistics | | 298.9 |
| 10 | n | 6 | 300.2 |
| 11 | xbar | 299.03 | 297.0 |
| 12 | Calculations | | |
| 13 | SE | 1.225 | |
| 14 | z | -0.789 | |
| 15 | Lower Test | | |
| 16 | lower_z | -1.645 | |
| 17 | Decision | Do Not Reject H0 | |
| 18 | Pvalue | 0.215 | |

Figure 6.4: Significance Test for a Normal Mean – Values

Is this convincing evidence that the mean content of cola bottles is less than the advertised 300 ml?

**Solution.** The alternative is $H_a : \mu < \mu_0$. Create the workbook template whose formulas are shown in Figure 6.3. Cells A14:A17, A18:A21, and A22:A25 respectively contain the calculations for lower, upper, and two-sided tests, respectively. To apply it here for a two-sided test, select rows 18–26, then choose **Format – Row – Hide** from the Menu Bar to hide rows 18–26 inclusive. Enter the data in B6:B11 and **Name** the range. Adjust the width of Column C. **Name** the other variables in the workbook. Enter the values sigma = "3," alpha = "0.05," null hypothesis = "300.0," and "lower" alternate hypothesis. The latter is only there to remind you what the test is. The Excel output is shown in Figure 6.4 which gives the critical value of $-z^* = -1.645$ at the 5% level of significance. The output also provides the $P$-value 0.215 and the conclusion not to reject $H_0$.

## Explanation

We encountered the function `NORMSINV` previously. The formula `= NORMSDIST` returns the cumulative normal distribution function $\Phi(z)$. For a one-sided lower test the $P$-value is the area to the left of the computed $z$ score $\frac{\bar{x}-\mu_0}{\sigma/\sqrt{n}}$ and is thus given by `= NORMSDIST`($z$). For an upper test, the $P$-value being the area to the right of $\frac{\bar{x}-\mu_0}{\sigma/\sqrt{n}}$ is given by `= 1 – NORMSDIST`($z$). For a two-sided test the formula for the $P$-value is `= 2*(1 – NORMSDIST`($|z|$)). The formula `= ABS`($z$) returns the absolute value of $z$. The decision rule uses the logical `= IF`(statement, true, false) which returns the string designated as true if the statement is true, or else it returns false.

# Chapter 7

# Inference for Distributions

In this chapter we develop Excel tools for constructing confidence intervals and carrying out significance tests for the mean of a single population and for comparing the means of two populations. These methods are based on the Student $t$-distribution.

## 7.1 Inference for the Mean of a Population

The workbooks in the previous chapter when $\sigma$ is known, which are based on the standardized score or normal distribution, will now be modified using the "Studentized" score, also called the Student $t$-distribution

$$\bar{x} \pm t^* \frac{s}{\sqrt{n}}$$

As before $n, \bar{x}$, and $s$ are the sample size, sample mean, and sample standard deviation, and $t^*$ is the critical $t$ value such that the area between $-t^*$ and $t^*$ under the curve of a $t$-density with $n-1$ degrees of freedom equals a specified $C$.

The Excel formula required is

$$= \texttt{TINV}(\alpha, \nu)$$

which returns the critical value for a level $C = 1 - \alpha$ confidence interval based on a $t$-distribution with $\nu$ degrees of freedom.

## One-Sample $t$ Confidence Intervals

**Example 7.1.** (Example 7.1 page 370 in text.) To study the metabolism of insects, researchers fed cockroaches measured amounts of a sugar solution. After 2, 5, and 10 hours, they dissected some of the cockroaches and measured the amount of sugar in various tissues.

Five roaches fed the sugar D-glucose and dissected after 10 hours had the following amounts (in micrograms) of D-glucose in their hindguts:

55.95 68.24 52.73 21.50 23.78

Find a 95% confidence interval for the mean amount of D-glucose in cockroach hindguts under these conditions.

| | A | B | C | D |
|---|---|---|---|---|
| 1 | *T Interval for a Normal Mean* | | | |
| 2 | | | | |
| 3 | | | | |
| 4 | User Input | | | |
| 5 | conf | 0.95 | | 0.95 |
| 6 | Summary Statistics | | | |
| 7 | n | 5 | | =COUNT(Data) |
| 8 | xbar | 44.4 | | =AVERAGE(Data) |
| 9 | Calculations | | | |
| 10 | s | 20.74 | | =STDEV(Data) |
| 11 | SE | 9.28 | | =s/SQRT(n) |
| 12 | df | 4 | | =n-1 |
| 13 | t | 2.776 | | =TINV(1-conf, df) |
| 14 | ME | 25.75 | | =t*SE |
| 15 | Confidence Limits | | | |
| 16 | lower | 18.69 | | =xbar-ME |
| 17 | upper | 70.19 | | =xbar+ME |
| 18 | | | | |
| 19 | Data | | | |
| 20 | 55.95 | | | |
| 21 | 68.24 | | | |
| 22 | 52.73 | | | |
| 23 | 21.50 | | | |
| 24 | 23.78 | | | |

Figure 7.1: Confidence Interval for a Normal Mean

**Solution.** Create the workbook shown in Figure 7.1 using steps analogous to the production of Figure 6.1. In place of an assumed standard deviation $\sigma$ we calculate the sample standard deviation $s$ using the Excel formula = STDEV(Data). The critical value $t^*$ (denoted by $t$ on the workbook) is obtained from the formula = TINV$(1 - conf, df)$ where $conf$ is the confidence level and $df = n - 1$ are the degrees of freedom. With this template in hand it is only necessary to input and name the data. We have typed the five data points in cells A20–A24 and **Named** them "Data," in cell A19. Column D has the formulas which you must enter in the respective cells in column B.

The output provided by Excel also appears in Figure 7.1 in addition to the data range. The 95% confidence interval for $\mu$ is (18.69, 70.19) from cells B16:B17.

## One-Sample $t$ Test

The workbook based on the normal distribution in the previous chapter is easily modified for use with the Student $t$. For the significance test of the null hypothesis

$$H_0 : \mu = \mu_0$$

the test statistic is

$$t = \frac{\bar{x} - \mu_0}{s/\sqrt{n}}$$

which has a Student $t$ distribution on $n - 1$ degrees of freedom. We remind the reader of the *unusual* Excel definition for the function which calculates the cumulative $t$, namely

$$\texttt{TDIST}(x, \nu, 1) = P[t(\nu) > x]$$

where $t(\nu)$ is a $t$-distribution on $\nu$ degrees of freedom. The argument $x$ must be positive and this accounts for the more complicated syntax in the decision rule in the corresponding template (Figure 7.2).

| | A | B | C | D | E |
|---|---|---|---|---|---|
| 1 | | *T Test for a Normal Mean* | | | |
| 2 | | | | | |
| 3 | | | | | |
| 4 | User Input | | | | |
| 5 | | null | | *Data* | |
| 6 | | alpha | | 2.0 | |
| 7 | | | | 0.4 | |
| 8 | Summary Stats | | | 0.7 | |
| 9 | | n | =COUNT(Data) | 2.0 | |
| 10 | | xbar | =AVERAGE(Data) | -0.4 | |
| 11 | Calculations | | | 2.2 | |
| 12 | | s | =STDEV(Data) | -1.3 | |
| 13 | | SE | =s/SQRT(n) | 1.2 | |
| 14 | | t | =(xbar-null)/SE | 1.1 | |
| 15 | | df | =n-1 | 2.3 | |
| 16 | Lower Alternative | | | | |
| 17 | | lower_t | =-TINV(2*alpha, df) | | |
| 18 | | Decision | =IF(t<lower_t,"Reject H0", "Do Not Reject H0") | | |
| 19 | | Pvalue | =IF(t<0, TDIST(ABS(t),df,1), 1-TDIST(t,df,1)) | | |
| 20 | Upper Alternative | | | | |
| 21 | | upper_t | =TINV(2*alpha, df) | | |
| 22 | | Decision | =IF(t>upper_t,"Reject H0","Do Not Reject H0") | | |
| 23 | | Pvalue | =IF(t>0, TDIST(t,df,1), 1-TDIST(ABS(t),df,1)) | | |
| 24 | Two-Sided Alternative | | | | |
| 25 | | two_t | =TINV(alpha, df) | | |
| 26 | | Decision | =IF(ABS(t)>two_t,"Reject H0","Do Not Reject H0") | | |
| 27 | | Pvalue | =TDIST(ABS(t), df, 2) | | |

Figure 7.2: One-Sample Student $t$-Test – Formulas

**Example 7.2.** (Example 7.2 page 371 in text.) Cola makers test new recipes for loss of sweetness during storage. Trained tasters rate the sweetness before and after storage. Here are the sweetness losses (sweetness before storage minus sweetness after storage) found by 10 tasters for one new cola recipe.

$$2.0 \quad 0.4 \quad 0.7 \quad 2.0 \quad -0.4 \quad 2.2 \quad -1.3 \quad 1.2 \quad 1.1 \quad 2.3$$

Are these data good evidence that the cola lost sweetness? Carry out the following test at the 1% level of significance. Also calculate the $P$-value.

$$\begin{aligned} H_0 : \mu &= 0 \\ H_a : \mu &> 0 \end{aligned}$$

**Solution.** Figure 7.2 gives the formulas for carrying out a $t$-test on a population mean. The user inputs are the null value 40, $\alpha = 0.01$, and the data. We have entered the data in cells D6:D15 shown on the Excel output in Figure 7.3 where we have hidden the rows of the template referring to two-sided or lower-tailed alternatives. The calculated $t$ value is 2.697 which is smaller than the 0.01 critical value for an upper-tailed test, namely 2.821. Therefore, the conclusion is to not reject $H_0$ at the 1% level of significance.

The $P$-value is provided in cell B23. Its value is 0.012 which still indicates strong evidence for a loss of sweetness although not at the 1% level.

| | A | B | C | D |
|---|---|---|---|---|
| 1 | *T Test for a Normal Mean* | | | |
| 2 | | | | |
| 3 | | | | |
| 4 | User Input | | | |
| 5 | null | 0.0 | | *Data* |
| 6 | alpha | 0.01 | | 2.0 |
| 7 | | | | 0.4 |
| 8 | Summary Stats | | | 0.7 |
| 9 | n | 10 | | 2.0 |
| 10 | xbar | 1.02 | | -0.4 |
| 11 | Calculations | | | 2.2 |
| 12 | s | 1.196 | | -1.3 |
| 13 | SE | 0.378 | | 1.2 |
| 14 | t | 2.697 | | 1.1 |
| 15 | df | 9 | | 2.3 |
| 20 | Upper Alternative | | | |
| 21 | upper_t | 2.821 | | |
| 22 | Decision | Do Not Reject H0 | | |
| 23 | Pvalue | 0.012 | | |

Figure 7.3: One-Sample Student $t$-Test – Values

## Matched Pairs $t$ Procedure

In order to reduce variability in a data set scientists sometimes use paired data matched on characteristics believed to affect the response. This is equivalent to a randomized block design. Such data are best analyzed if one-sample procedures are applied to differences between the pairs. There is a loss in degrees of freedom for error but if the matching is effective then this will be more than offset by the gain in reduced variance of the differences. The same method applies to before-after measurements on the same subjects.

Thus, we may apply the workbooks in Chapter 6 to the differences. Excel also provides a direct method for the matched pairs $t$-test using the **Analysis ToolPak**. We describe both approaches applied to the same data set.

**Example 7.3.** (Example 7.3 page 375 in text.) We hear that listening to Mozart improves students' performance on tests. Perhaps pleasant odors have a similar effect. To test this idea, 21 subjects worked a paper-and-pencil maze while wearing a mask. The mask was either unscented or carried a floral scent. The response variable is their mean time on three trials. Each subject worked the maze with both masks, but in random order. The randomization is important because subjects tend to improve their times as they work a maze repeatedly. Table 7.1 gives the subjects' times.

Table 7.1: Mean Time to Complete a Maze

| Subject | Unscented | Scented | Subject | Unscented | Scented |
|---|---|---|---|---|---|
| 1 | 30.60 | 37.97 | 12 | 58.93 | 83.50 |
| 2 | 48.43 | 51.57 | 13 | 54.47 | 38.30 |
| 3 | 60.77 | 56.67 | 14 | 43.53 | 51.37 |
| 4 | 36.07 | 40.47 | 15 | 37.93 | 29.33 |
| 5 | 68.47 | 49.00 | 16 | 43.50 | 54.27 |
| 6 | 32.43 | 43.23 | 17 | 87.70 | 62.73 |
| 7 | 43.70 | 44.57 | 18 | 53.53 | 58.00 |
| 8 | 37.10 | 28.40 | 19 | 64.30 | 52.40 |
| 9 | 31.17 | 28.23 | 20 | 47.37 | 53.63 |
| 10 | 51.23 | 68.47 | 21 | 53.67 | 47.00 |
| 11 | 65.40 | 51.10 | | | |

To analyze these data subtract the scented time from the unscented time for each subject. The 21 differences form a single sample. Because shorter times represent better performance, positive differences show that the subject did better when wearing the scented mask. To assess whether the floral scent significantly improved performance, we test

$$\begin{aligned} H_0 : \mu &= 0 \\ H_a : \mu &> 0 \end{aligned}$$

| | A | B | C | D | E | F |
|---|---|---|---|---|---|---|
| 1 | *Paired T Test for a Normal Mean (Direct Approach)* | | | | | |
| 2 | | | | | | |
| 3 | | | | *Unscented* | *Scented* | *Difference* |
| 4 | User Input | | | 30.60 | 37.97 | -7.37 |
| 5 | null | 0.00 | | 48.43 | 51.57 | -3.14 |
| 6 | alpha | 0.05 | | 60.77 | 56.67 | 4.10 |
| 7 | | | | 36.07 | 40.47 | -4.40 |
| 8 | Summary Stats | | | 68.47 | 49.00 | 19.47 |
| 9 | n | 21 | | 32.43 | 43.23 | -10.80 |
| 10 | xbar | 0.957 | | 43.70 | 44.57 | -0.87 |
| 11 | Calculations | | | 37.10 | 28.40 | 8.70 |
| 12 | s | 12.548 | | 31.17 | 28.23 | 2.94 |
| 13 | SE | 2.738 | | 51.23 | 68.47 | -17.24 |
| 14 | t | 0.349 | | 65.40 | 51.10 | 14.30 |
| 15 | df | 20 | | 58.93 | 83.50 | -24.57 |
| 16 | Upper Alternative | | | 54.47 | 38.30 | 16.17 |
| 17 | upper_t | 1.725 | | 43.53 | 51.37 | -7.84 |
| 18 | Decision | Do Not Reject H0 | | 37.93 | 29.33 | 8.60 |
| 19 | Pvalue | 0.365 | | 43.50 | 54.27 | -10.77 |
| 20 | | | | 87.70 | 62.73 | 24.97 |
| 21 | | | | 53.53 | 58.00 | -4.47 |
| 22 | | | | 64.30 | 52.40 | 11.90 |
| 23 | | | | 47.37 | 53.63 | -6.26 |
| 24 | | | | 53.67 | 47.00 | 6.67 |

Figure 7.4: Paired $t$-Test – Direct Approach

Here $\mu$ is the mean difference (unscented – scented) in the population from which the subjects were drawn. The null hypothesis says that no improvement occurs, and the alternative says that the unscented times are longer than the scented times on the average.

## Matched Pairs $t$ Test – Direct Approach

1. Open a new workbook using the template shown in Figure 7.2 but delete rows 16–19 and 24–27 to customize for an upper test (Figure 7.4).

2. It only remains to enter "0" for null, "0.05" for alpha, and record the data in a convenient place. In Figure 7.4 we have used columns D, E, and F for the data. Enter the label "Unscented " in cell D3 followed by the unscented data in cells D4:D24. Enter the label "Scented " in cell E3 followed by the scented data in cells E4:E24. Enter the label "Difference " in cell F3. Then in cell F4 enter the difference D4–E4 in the **Formula Bar**. Select cell F4 and then move the mouse pointer over the fill handle in the lower right corner of cell F4. The pointer changes from an outline plus sign to a cross hair + when you are in position on the fill handle. Click the fill handle and drag it down so that the range F4:F24 is selected. Release the mouse button and Excel fills the contents of cells F4:F24 with the corresponding differences unscented – scented.

3. **Name** the difference range by selecting F3:F24, and from the Menu Bar

| | A | B | C | D | E | F |
|---|---|---|---|---|---|---|
| 1 | | | *T Test for a Normal Mean (Using the Analysis ToolPak)* | | | |
| 2 | | | | | | |
| 3 | Unscented | Scented | Difference | t-Test: Paired Two Sample for Means | | |
| 4 | 30.60 | 37.97 | -7.37 | | | |
| 5 | 48.43 | 51.57 | -3.14 | | *Unscented* | *Scented* |
| 6 | 60.77 | 56.67 | 4.10 | Mean | 50.01429 | 49.05762 |
| 7 | 36.07 | 40.47 | -4.40 | Variance | 206.3097 | 179.1748 |
| 8 | 68.47 | 49.00 | 19.47 | Observations | 21 | 21 |
| 9 | 32.43 | 43.23 | -10.80 | Pearson Correlation | 0.593026 | |
| 10 | 43.70 | 44.57 | -0.87 | Hypothesized Mean Difference | 0 | |
| 11 | 37.10 | 28.40 | 8.70 | df | 20 | |
| 12 | 31.17 | 28.23 | 2.94 | t Stat | 0.349381 | |
| 13 | 51.23 | 68.47 | -17.24 | P(T<=t) one-tail | 0.365227 | |
| 14 | 65.40 | 51.10 | 14.30 | t Critical one-tail | 1.724718 | |
| 15 | 58.93 | 83.50 | -24.57 | P(T<=t) two-tail | 0.730455 | |
| 16 | 54.47 | 38.30 | 16.17 | t Critical two-tail | 2.085962 | |
| 17 | 43.53 | 51.37 | -7.84 | | | |
| 18 | 37.93 | 29.33 | 8.60 | | | |
| 19 | 43.50 | 54.27 | -10.77 | | | |
| 20 | 87.70 | 62.73 | 24.97 | | | |
| 21 | 53.53 | 58.00 | -4.47 | | | |
| 22 | 64.30 | 52.40 | 11.90 | | | |
| 23 | 47.37 | 53.63 | -6.26 | | | |
| 24 | 53.67 | 47.00 | 6.67 | | | |

Figure 7.5: Paired $t$-Test – Analysis ToolPak

choose **Insert – Name – Create** and then check the Top Row box in the **Create Names** dialog box. We can now refer to the data in F4:F24 by the name "Difference." (This is convenient but not necessary; we could equally use the cell reference F4:F24.)

4. Finally, where the name "Data" is used in formulas in the previous template of Figure 7.2, replace it with the name "Difference."

### Excel Output

Figure 7.4 gives the Excel output. The computed $t$-statistic is 0.349 which is less than the 5% critical value of 1.725. We therefore do not reject $H_0$. Additionally we find that the $P$-value is 0.365.

## Matched Pairs $t$ Test – Using the ToolPak

Excel also provides an **Analysis** tool for a matched pairs $t$ test. However, it cannot be used with summarized data while the direct approach can.

1. Open a new workbook and enter the unscented and scented values with their labels in cells A3:A24 and B3:B24, respectively as in Figure 7.5.

2. In cell C3 enter the label "Difference." In cell C4 type "$= A4 - B4$." Then select C4 and fill to cells C4:C24.

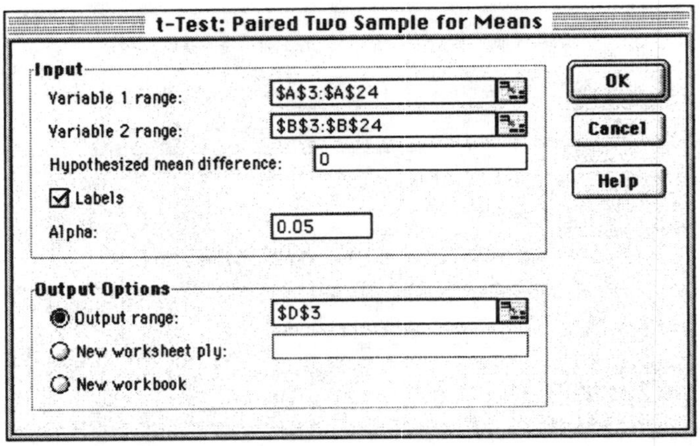

t-Test: Paired Two Sample for Means

Input
Variable 1 range: $A$3:$A$24
Variable 2 range: $B$3:$B$24
Hypothesized mean difference: 0
☑ Labels
Alpha: 0.05

Output Options
◉ Output range: $D$3
○ New worksheet ply:
○ New workbook

OK
Cancel
Help

Figure 7.6: Paired $t$-Test – Dialog Box

3. Choose **Tools – Data Analysis** from the Menu Bar, and then check the box ***t*-Test:Paired Two Sample for Means**. Click OK.

4. Complete the next dialog box as shown in Figure 7.6. **Caution** is required in determining which scores are entered for **Variable 1 range** and which scores are entered for **Variable 2 range**. Since $H_a : \mu > 0$ we are taking unscented–scented scores. Hence enter the unscented data range A3:A24 for Variable 1 and the scented data range B3:B24 for Variable 2.

### Excel Output

The output appears in Figure 7.5 in the range D3:F16 beginning with cell D3 as specified in Figure 7.6. Individual sample means are given as well as the value of the test statistic $t$ in cell E10. One-sided and two-sided critical values are provided in E12 and E14 as well as corresponding $P$-values in E11 and E13. Because $H_0 : \mu > 0$ and $t = 0.349$ we reject at level $\alpha = 0.01$. The $P$-value is 0.365 (which is half of the two-sided $P$-value given in cell E15).

The entry "$P(T <= t)$one-tail" in E13 is not $P(t(20) \leq 0.349)$ as the notation would seem to suggest. Rather it represents the tail area relative to $t$ (so it is a lower tail if $t$ is negative and an upper tail if $t$ is positive). It is therefore not always a $P$-value.

### Confidence Interval for Paired Data

The **Analysis ToolPak** does not provide a confidence interval directly but it provides the information needed to carry out the calculations using the template in Section 7.1 applied to the differences. Open the workbook that you used for Example 7.1 and edit the template, replacing the word "Data" in cells B7, B8, and B10 by pointing to cells C4:C24 in the workbook you used for Example 7.3, shown in Figure 7.5). Excel will take the values from one workbook and use them in another. This is a reminder that data need not be on the same sheet or even the same workbook as your analysis.

> **Exercise.** For Example 7.3, show that a 90% confidence interval for the mean of the difference between unscented and scented scores is
>
> $$(-3.766, 5.679)$$

## 7.2 Comparing Two Means

Independent SRS of sizes $n_1$ and $n_2$ are obtained from populations with means $\mu_1$ and $\mu_2$, respectively. We are interested in comparing $\mu_1$ and $\mu_2$. The appropriate statistics and critical values required depend on the assumptions made. Suppose that the data is collected from normal populations with standard deviations $\sigma_1, \sigma_2$. Denote by $\bar{x}_1, s_1^2, \bar{x}_2$, and $s_2^2$ the corresponding summary statistics (the sample means and sample variances).

In the following we assume that the underlying populations are normal. Inference is then based on a two-sample statistic of the form

$$\frac{(\bar{x}_1 - \bar{x}_2) - (\mu_1 - \mu_2)}{\text{SE}} \tag{7.1}$$

where SE represents the standard error of the numerator.

If the population standard deviations $\sigma_1$ and $\sigma_2$ are known then we use

$$z = \frac{(\bar{x}_1 - \bar{x}_2) - (\mu_1 - \mu_2)}{\sqrt{\frac{\sigma_1^2}{n_1} + \frac{\sigma_2^2}{n_2}}}$$

which has a standard normal distribution.

If $\sigma_1$ and $\sigma_2$ are unknown then the appropriate ratio is

$$t = \frac{(\bar{x}_1 - \bar{x}_2) - (\mu_1 - \mu_2)}{\sqrt{\frac{s_1^2}{n_1} + \frac{s_2^2}{n_2}}} \tag{7.2}$$

which is called a two-sample $t$ statistic.

Actually, the distribution of the statistic in (7.2) depends on $\sigma_1$ and $\sigma_2$ and does not have an exact $t$-distribution. Nonetheless, it is used with $t$ critical values in inference in one of two ways, each involving a computed value for a degrees of freedom $\nu$ associated with the denominator of (7.1) to provide an approximate $t$ statistic. These two options are:

1. Use a value for $\nu$ given by

$$\nu = \frac{\left(\frac{s_1^2}{n_1} + \frac{s_2^2}{n_2}\right)^2}{\frac{1}{n_1-1}\left(\frac{s_1^2}{n_1}\right)^2 + \frac{1}{n_2-1}\left(\frac{s_2^2}{n_2}\right)^2}$$

2. Use

$$\nu = \min\{n_1 - 1, n_2 - 1\}$$

Approximation (2) is conservative in the sense of providing a larger margin of error and is recommended in the text when doing calculations without the aid of software. Approximation (1) is considered to provide a quite accurate approximation to the actual distribution.

Excel, as well as most statistical software, uses (1) and provides three tools in the **Analysis ToolPak** for analyzing independent samples from normal populations: $z$-test; $t$-test (unequal variances); and pooled $t$-test (equal variances). These involve different sets of assumptions and consequently different forms for SE. Each tool provides a similar dialog box for user input, parameters, and the range for the actual raw data.

However, sometimes only summary statistics are available and the **Analysis ToolPak** cannot be used; instead, direct calculations are required. The first example below deals with summarized data.

## Using Summarized Data

Suppose only $\bar{x}_1, s_1^2, \bar{x}_2$, and $s_2^2$, not the raw data, are available. We will use Excel to calculate (7.2) and use it for a test of significance and a confidence interval.

> **Example 7.4.** (Example 7.9 page 398 in text.) The Chapin Social Insight Test is a psychological test designed to measure how accurately a person appraises other people. The possible scores on the test range from 0 to 41. During the development of the Chapin test, it was given to several different groups of people. Here are the results for male and female college students majoring in the liberal arts.

Table 7.2: Chapin Social Insight Test

| Group | Sex | $n$ | $\bar{x}$ | $s$ |
|---|---|---|---|---|
| 1 | Male | 133 | 25.34 | 5.05 |
| 2 | Female | 162 | 24.94 | 5.44 |

Do these data support the contention that female and male students differ in average social insight?

| | A | B | C | D | F | G |
|---|---|---|---|---|---|---|
| 1 | | *Two-Sample T Test Summarized Data* | | | | |
| 2 | | | | | | |
| 3 | | Summary Statistics and User Input | | | | |
| 4 | Group | n | xbar | s | | |
| 5 | Male | 133 | 25.34 | 5.05 | | |
| 6 | Female | 162 | 24.94 | 5.44 | | |
| 7 | | | | | | |
| 8 | SE | 0.612 | =SQRT(D5^2/B5+D6^2/B6) | | | |
| 9 | t | 0.654 | =((C5-C6)-null)/SE | | | |
| 10 | alpha | 0.05 | | | | |
| 11 | null | 0 | | | | |
| 12 | mindf | 132 | =MIN(B5-1,B6-1) | | | |
| 13 | numerator | 0.140 | =POWER((D5^2/B5+D6^2/B6),2) | | | |
| 14 | denominator | 0.000486 | =(D5^4/B5^2)/(B5-1)+(D6^4/B6^2)/(B6-1) | | | |
| 15 | df | 289 | =1+INT(B13/B14) | | | |
| 16 | Lower Test | | | | | |
| 17 | lower_t | | =-TINV(2*alpha, mindf) | | | |
| 18 | Decision | | =IF(t<lower_t,"Reject H0", "Do Not Reject H0") | | | |
| 19 | Pvalue | | =IF(t<0, TDIST(ABS(t),mindf,1), 1-TDIST(t,mindf,1)) | | | |
| 20 | Upper Test | | | | | |
| 21 | upper_t | | =TINV(2*alpha, mindf) | | | |
| 22 | Decision | | =IF(t>upper_t,"Reject H0","Do Not Reject H0") | | | |
| 23 | Pvalue | | =IF(t>0, TDIST(t,mindf,1), 1-TDIST(ABS(t),mindf,1)) | | | |
| 24 | Two-Sided Test | | | | | |
| 25 | two_t | 1.978 | =TINV(alpha, mindf) | | | |
| 26 | Decision | Do Not Reject H0 | =IF(ABS(t)>two_t,"Reject H0 ',"Do Not Reject H0") | | | |
| 27 | Pvalue | 0.5144 | =TDIST(ABS(t), mindf, 2) | | | |

Figure 7.7: Two-Sample $t$ Test – Summarized Data

**Solution.** The hypothesis to be tested is

$$\begin{aligned} H_0 : \mu_1 &= \mu_2, \\ H_a : \mu_1 &\neq \mu_2 \end{aligned}$$

where $\mu_1$ and $\mu_2$ are the hypothesized mean scores on the test for a population of all male students and all female students, respectively.
At the level $\alpha = 0.05$ the decision rule is to reject $H_0$ if the computed $t$ value

$$t = \frac{\bar{x}_1 - \bar{x}_2}{\sqrt{\frac{s_1^2}{n_1} + \frac{s_2^2}{n_2}}}$$

satisfies $|t| > t^*$ where $t^*$ is the upper $\alpha/2 = 0.025$ critical value of a $t$ distribution with degrees of freedom $132 = min\{133 - 1, 162 - 1\}$. The calculations can readily be made on a hand calculator and compared with critical values from a $t$-table.

While the full power of Excel comes from dealing with large data sets, even this simple example can illustrate the use of a spreadsheet to evaluate formulas within equations. Figure 7.7 is an Excel workbook containing a template for summarized data. For the data in this problem, the calculations are carried out in column B. The actual formulas in the cells behind column B are given in the adjacent cells in column C. Although this is a two-sided test, formulas are also provided for upper and lower tests, only one of which should be used at any time. We have used

**Named Ranges** rather than cell references for purposes of clarity. Remember to name your ranges when copying and adapting this template to your own workbook.

From Figure 7.7 we can read the computed $t = 0.654$ in cell B9. The critical $t$ value is 1.978 from cell B25, and the $P$-value is 0.5144 in cell B27. The conclusion, given in cell B26, is that the data give no evidence of a male/female difference in mean social insight score.

The more accurate value for the degrees of freedom $\nu = 289$ is also provided in Figure 7.7 and can be used if the variable *mindf* is replaced by *df* in the formulas in the template. In view of the large sample size, the computed $t$ is close to a standard normal $z$, the critical value is close to the corresponding normal critical value 1.96, and likewise the $P$-value is approximated with

$$P\text{-value } = 2\,(1 - \Phi(|z|))$$

where $\Phi$ is the cumulative $N(0,1)$ distribution function. While the above template has been designed for summarized data, it can easily be modified for use with raw data, for instance, by entering `= AVERAGE`(Range) and `= STDEV`(Range) in place of the *values* of the corresponding sample means or standard deviations, where Range is the cell range for the samples. Excel computes the sample means and standard deviations from the data and inserts them where required in the formulas.

However, with raw data it is preferable to use Excel's **Analysis ToolPak**.

## Using the Analysis ToolPak

As mentioned, Excel provides three tools in the **Analysis ToolPak** for comparing means from two populations based on independent samples. These are the two-sample $t$, pooled two-sample $t$, and the two-sample $Z$ tests. These tools provide dialog boxes in which the user locates the data and decides on the type of analysis desired. For large sample sizes, the results of two-sample $t$ and $Z$ options are virtually the same and the sequence of steps in the two tools are identical. Therefore, as we will be discussing the two-sample $t$ ToolPak in detail we will not illustrate its $Z$ counterpart. Besides, there is a bug in the two-sample $Z$ ToolPak which outputs an incorrect two-sided $P$-value, and moreover, there is no built-in confidence interval procedure. These considerations limit the usefulness of the two-sample $Z$ ToolPak whose use is not recommended.

**Example 7.5.** (Exercise 7.37 page 404 in text.) In a randomized experiment, researchers compared 6 white rats poisoned with DDT with a control group of 6 nonpoisoned rats. Electrical measurements of nerve activity are the main clue to the nature of DDT poisoning. When a nerve is stimulated, its electrical response shows a sharp spike followed by a much smaller second spike. The experiment found that the second spike is larger in rats fed DDT than in normal rats. The researchers measured the height of the second spike as a percent of the first spike when a nerve in the rat's leg was stimulated. The results of the experiment were:

| Poisoned | 12.207 | 16.869 | 25.050 | 22.429 | 8.456 | 20.589 |
|---|---|---|---|---|---|---|
| Control | 11.074 | 9.686 | 12.064 | 9.351 | 8.182 | 6.642 |

The researchers wondered if poisoned rats differed from unpoisoned rats. Carry out an appropriate test of significance.

**Solution.** Denote by $\mu_1$ and $\mu_2$ the mean percentages for hypothetical populations of all possible poisoned and unpoisoned rats, respectively. The significance test to be carried out is:

$$\begin{aligned} H_0 : \mu_1 &= \mu_2 \\ H_a : \mu_1 &\neq \mu_2 \end{aligned}$$

1. Open a new workbook and enter the above data. Insert the poisoned group in cells A3:A8 and the control group in cells B3:B8. Enter the label "Poisoned" in A2 and the label "Control" in B2. (See Figure 7.10 showing output.)

2. From the Menu Bar choose **Tools – Data Analysis** and select ***t*-Test: Two-Sample Assuming Unequal Variances** from the list of selections (Figure 7.8). Click OK (equivalently, double-click your selection).

3. A dialog box (Figure 7.9) appears. Complete as shown. **Variable 1 range** refers to the cell addresses of the sample you have designated by subscript 1, in this case the Poisoned group. Type A2:A8 in its text area (with the flashing vertical I-beam). Alternatively, you can point to the data by clicking on cell A2 and dragging to the end of cell A8. The values $A$2:$A$8 will appear in the text area of the dialog box. Similarly, enter the range B2:B8 for the Control group in the **Variable 2 range**.

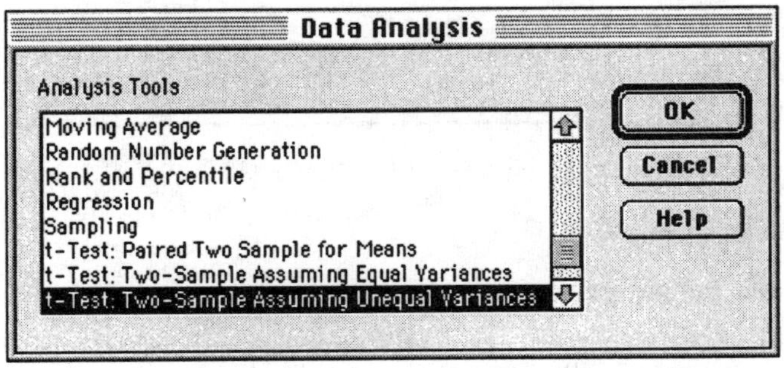

Figure 7.8: Two-Sample $t$ Analysis Tools

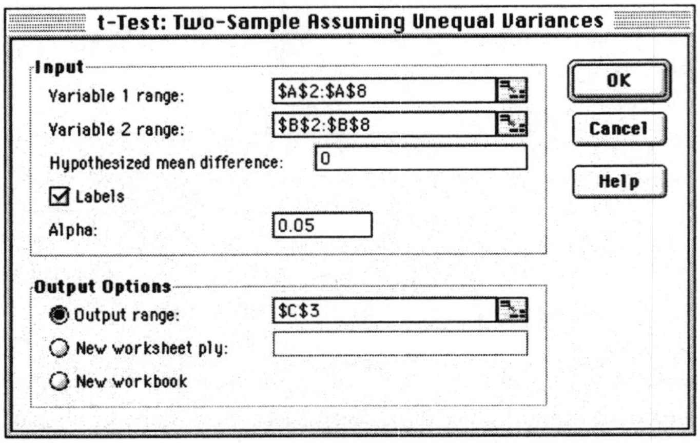

t-Test: Two-Sample Assuming Unequal Variances

Input

Variable 1 range: $A$2:$A$8

Variable 2 range: $B$2:$B$8

Hypothesized mean difference: 0

☑ Labels

Alpha: 0.05

Output Options

◉ Output range: $C$3

○ New worksheet ply:

○ New workbook

OK

Cancel

Help

Figure 7.9: Two-Sample $t$ Dialog Box

4. The **Hypothesized mean difference** refers to the null value, which is "0" here. Check the **Labels** box because your ranges included the labels for the two groups. The level of significance **Alpha** is the default "0.05." We will be placing the output in the same workbook as the data, so check the radio button **Output range** and type C3 in the text area. Finally, click OK. The output will appear in a block of cells whose upper left corner is cell C3.

| | A | B | C | D | E | F |
|---|---|---|---|---|---|---|
| 1 | | | *Two-Sample t Test* | | | |
| 2 | Poisoned | Control | | | | |
| 3 | 12.207 | 11.074 | t-Test: Two-Sample Assuming Unequal Variances | | | |
| 4 | 16.869 | 9.686 | | | | |
| 5 | 25.050 | 12.064 | | *Poisoned* | *Control* | |
| 6 | 22.429 | 9.351 | Mean | 17.6 | 9.499833 | |
| 7 | 8.456 | 8.182 | Variance | 40.197482 | 3.802731 | |
| 8 | 20.589 | 6.642 | Observations | 6 | 6 | |
| 9 | | | Hypothesized Mean Difference | 0 | | |
| 10 | | | df | 6 | | |
| 11 | | | t Stat | 2.9911775 | | |
| 12 | | | P(T<=t) one-tail | 0.0121416 | | |
| 13 | | | t Critical one-tail | 1.9431809 | | |
| 14 | | | P(T<=t) two-tail | 0.0242832 | | |
| 15 | | | t Critical two-tail | 2.4469136 | | |
| 16 | | | | | | |
| 17 | | | Mean Difference | 8.100 | =D6-E6 | |
| 18 | | | SE | 2.708 | =SQRT(D7/D8+E7/E8) | |
| 19 | | | critical t | 1.943 | =TINV(0.10, D10) | |
| 20 | | | ME | 5.26 | =D19*D18 | |
| 21 | | | lower | 2.84 | =D17-D20 | |
| 22 | | | upper | 13.36 | =D17+D20 | |

Figure 7.10: Two-Sample $t$ ToolPak Output

## Excel Output

The output appears in the range C3:E15 in Figure 7.10. The range C17:E22 is not part of the output but is the result of additional formulas we have entered to give confidence intervals (see below). From cells D6:E6 we see that the sample means for poisoned and control groups are 17.600 and 9.450, while the sample variances are 40.197 and 3.803, respectively. The degrees of freedom are 6 (Excel rounds up to the nearest integer). The computed $t$ statistic in cell D11 is 2.991 while the two-sided critical $t^*$ value on 6 degrees of freedom at the 5% level in cell D15 is 2.447. Since the computed $t$ exceeds $t^*$ we reject the null hypothesis and conclude that there is good evidence that the mean size of the secondary spike is different (larger, in fact) in rats fed DDT.

Excel provides $P$-values in cells D12 and D14. As before, the entry "$P(T <= t)$ one-tail" in C12 needs some explanation. It is meant to be a one-tailed $P$-value which depends on the calculated $t$ stat, the computed $t$ statistic. If $t$ stat $< 0$, then $P(T <= t)$ one-tail is in fact the lower tail corresponding to the area to the left of $t$ stat under a $t$ density curve. But if $t$ is positive then $P(T <= t)$ one-tail is the area to the right of $t$ stat. It is therefore not always the $P$-value. For instance, if the test to be carried out were

$$\begin{aligned} H_0 : \mu_1 &= \mu_2 \\ H_a : \mu_1 &< \mu_2 \end{aligned}$$

then the $P$-value would be $1 - 0.0121 = 0.9879$ rather than 0.0121. $P(T <= t)$ two-tail is the correct $P$-value for a two-tailed test so we can also conclude that the $P$-value is 0.024.

## Confidence Intervals

The **ToolPak** does not print a confidence interval directly, but the output provides enough information to carry out the calculations. Details are given in cells C17:E22 of Figure 7.10, and are also shown isolated in Figure 7.11. The cells in column E of this block show the formulas which are the entries behind the cells in column D and whose values are evaluated and printed in the workbook by Excel. These formulas are the Excel equivalents of the formula

$$\bar{x}_1 - \bar{x}_2 \pm t^* \sqrt{\frac{s_1^2}{n_1} + \frac{s_2^2}{n_2}}$$

The information needed – the sample means, sample variances, sample sizes, and the critical $t^*$ values – are part of the **ToolPak** output and is referenced in cells C17:E22.

**Example 7.5 (continued).** Find a 90% confidence interval for the mean difference between poisoned and control rats.

| | C | D | E |
|---|---|---|---|
| 17 | Mean Difference | 8.100 | =D6-E6 |
| 18 | SE | 2.708 | =SQRT(D7/D8+E7/E8) |
| 19 | critical t | 1.943 | =TINV(0.10, D10) |
| 20 | ME | 5.26 | =D19*D18 |
| 21 | lower | 2.84 | =D17-D20 |
| 22 | upper | 13.36 | =D17+D20 |

Figure 7.11: Two-Sample $t$ Confidence Interval Formulas

**Solution.** From cells D21:D22 of Figure 7.11 we can read the confidence interval

$$(2.84, 13.36)$$

## Pooled Two-Sample $t$ Procedures

When the two populations are believed to be normal with the same variance, it is more common to use a pooled two-sample $t$ based on an exact $t$-distribution. The procedures are based on the statistic

$$t = \frac{(\bar{x}_1 - \bar{x}_2) - (\mu_1 - \mu_2)}{s_p\sqrt{\frac{1}{n_1} + \frac{1}{n_2}}} \tag{7.3}$$

where $s_p^2 = \frac{(n_1-1)s_1^2+(n_2-1)s_2^2}{n_1+n_2-2}$ is called the pooled sample variance. This statistic is known to have an exact $t$ distribution on $\nu = n_1 + n_2 - 2$ degrees of freedom. The previous two-sample $t$ analyses carry over with the obvious modifications for the degrees of freedom and use of the denominator of (7.3) in place of the denominator of (7.2).

**Example 7.6.** (Exercise 7.64 page 422 in text.) Nitrites are often added to meat products as preservatives. In a study of the effect of nitrites on bacteria, researchers measured the rate of uptake of an amino acid for 60 cultures of bacteria: 30 growing in a medium to which nitrites had been added and another 30 growing in a standard medium as a control group. Table 7.7 in your text gives the data from this study. Carry out a test of the research hypothesis that nitrites decrease amino acid uptake and report your results.

**Solution.** The significance test is

$$\begin{aligned} H_0 : \mu_1 &= \mu_2 \\ H_a : \mu_1 &> \mu_2 \end{aligned}$$

where $\mu_1$ represents the mean amino acid uptake by the control group and $\mu_2$ represents the mean amino acid uptake by the nitrite group.

1. Enter the above data and labels in cells A3:A33 (Control) and cells B3:B33 (Nitrite) of a workbook.

2. From the Menu Bar choose **Tools** – **Data Analysis** and select ***t*-Test: Two-Sample Assuming Equal Variances** from the list of selections. (Refer to the dialog box in Figure 7.8.) Click OK.

3. Complete the next dialog box exactly as you did for the unequal variances case.

## Excel Output

The output appears in the range C3:E16 in Figure 7.12. As the entire data set is in the CD accompanying the text, we have only reproduced the relevant part of the workbook with only 20 observations from each group. We observe that

$$\begin{aligned} \bar{x}_1 &= 8.073 & s_1^2 &= 1.141 \\ \bar{x}_2 &= 7.807 & s_2^2 &= 1.526 \\ s_p^2 &= 1.334 \end{aligned}$$

| | A | B | C | D | E | F |
|---|---|---|---|---|---|---|
| 1 | | | | *Pooled Two-Sample t Test* | | |
| 2 | | | | | | |
| 3 | Control | Nitrite | t-Test: Two-Sample Assuming Equal Variances | | | |
| 4 | 6.450 | 8.303 | | | | |
| 5 | 9.011 | 8.534 | | *Control* | *Nitrite* | |
| 6 | 7.821 | 7.688 | Mean | 8.07293 | 7.8073 | |
| 7 | 6.579 | 8.568 | Variance | 1.14132 | 1.52571 | |
| 8 | 8.066 | 8.100 | Observations | 30 | 30 | |
| 9 | 6.679 | 8.040 | Pooled Variance | 1.33351 | | |
| 10 | 9.032 | 5.589 | Hypothesized Mean Difference | 0 | | |
| 11 | 7.061 | 6.529 | df | 58 | | |
| 12 | 8.368 | 8.106 | t Stat | 0.8909 | | |
| 13 | 7.238 | 7.901 | P(T<=t) one-tail | 0.18833 | | |
| 14 | 8.709 | 8.252 | t Critical one-tail | 1.67155 | | |
| 15 | 9.036 | 10.227 | P(T<=t) two-tail | 0.37666 | | |
| 16 | 9.996 | 6.811 | t Critical two-tail | 2.00172 | | |
| 17 | 10.333 | 7.708 | | | | |
| 18 | 7.408 | 6.281 | Mean Difference | 0.266 | =D6-E6 | |
| 19 | 8.621 | 9.489 | SE | 0.298 | =SQRT(D9)*SQRT((1/D8 + 1/E8)) | |
| 20 | 7.128 | 9.460 | critical t | 1.672 | =TINV(0.1, D11) | |
| 21 | 8.128 | 6.201 | ME | 0.498 | =D20*D19 | |
| 22 | 8.516 | 4.972 | lower | -0.233 | =D18-D21 | |
| 23 | 8.830 | 8.226 | upper | 0.764 | =D18+D21 | |

Figure 7.12: Two-Sample $t$ ToolPak Output

The computed pooled $t$-statistic on 58 degrees of freedom is (cell D12)

$$t = 0.8909$$

The $\alpha = 0.05$ upper critical value is (cell D14)

$$t^* = 1.672$$

We conclude that there is no significant decrease in the amino acid uptake due to the nitrites. The $P$-value 0.377 is shown in cell D15.

### Confidence Intervals

In order to supplement the **Analysis ToolPak** output which does not provide a confidence interval we have also included in C18:E23 of Figure 7.12 formulas and output for the confidence intervals. These rely on the summary statistics produced by the ToolPak and are the Excel equivalents of the formula

$$\bar{x}_1 - \bar{x}_2 \pm t^* s_p \sqrt{\frac{1}{n_1} + \frac{1}{n_2}}$$

## 7.3 Inference for Population Spread

Suppose that $s_1^2$ and $s_2^2$ are the sample variances of independent simple random samples of sizes $n_1$ and $n_2$ taken from normal populations $N(\mu_1, \sigma_1)$ and $N(\mu_2, \sigma_2)$, respectively. Then the ratio

$$F = \frac{s_1^2/\sigma_1^2}{s_2^2/\sigma_2^2}$$

has a known sampling distribution which does not depend on $\{\mu_1, \mu_2, \sigma_1, \sigma_2\}$, but only on the sample sizes. It has an $F$-distribution on $n_1 - 1$ and $n_2 - 1$ degrees of freedom, for the numerator and the denominator, respectively. The ratio on the right side of the equation is only one manifestation of the $F$ distribution, which is also used in analysis of variance and regression.

### Two-Sample $F$ Test

In this section the context is comparison of $\sigma_1$ and $\sigma_2$. It turns out for mathematical reasons that the appropriate parameter for testing the null hypothesis

$$H_0 : \sigma_1 = \sigma_2$$

is the ratio $\frac{\sigma_1}{\sigma_2}$ (equivalently $\frac{\sigma_1^2}{\sigma_2^2}$) rather than the difference which we used for comparing means.

**Example 7.7.** (Based on data in Exercise 7.43 page 409 in text.) A study of computer-assisted learning examined the learning of "Blissymbols" by children. Blissymbols are pictographs (think of Egyptian hieroglyphics) that are sometimes used to help learning-impaired children communicate. The researcher designed two computer lessons that taught the same content using the same examples. One lesson required the children to interact with the material, while in the other, the children controlled only the pace of the lesson. Call these two styles "Active" and "Passive." After the lesson, the computer presented a quiz that asked the children to identify 56 Blissymbols. Following are the numbers of correct identifications by the 24 children in the Active group:

| | | | | | | | | | | | |
|---|---|---|---|---|---|---|---|---|---|---|---|
| 29 | 28 | 24 | 31 | 15 | 24 | 27 | 23 | 20 | 22 | 23 | 21 |
| 24 | 35 | 21 | 24 | 44 | 28 | 17 | 21 | 21 | 20 | 28 | 16 |

and the numbers of the 24 children in the Passive group:

| | | | | | | | | | | | |
|---|---|---|---|---|---|---|---|---|---|---|---|
| 16 | 14 | 17 | 15 | 26 | 17 | 12 | 25 | 21 | 20 | 18 | 21 |
| 20 | 16 | 18 | 15 | 26 | 15 | 13 | 17 | 21 | 19 | 15 | 12 |

Compare the variability between Active and Passive counts by carrying out a significance test of

$$\begin{aligned} H_0 : \sigma_1 &= \sigma_2 \\ H_a : \sigma_1 &\neq \sigma_2 \end{aligned}$$

using level $\alpha = 0.05$ where $\sigma_1$ and $\sigma_2$ are the standard deviations of the correct identification counts in the Active and Passive populations, respectively.

| | A | B | C | D | E | F |
|---|---|---|---|---|---|---|
| 1 | | | *F Test for Equality of Variances* | | | |
| 2 | | | | | | |
| 3 | Active | Passive | | | | |
| 4 | 29 | 16 | | F-Test Two-Sample for Variances | | |
| 5 | 28 | 14 | | | | |
| 6 | 24 | 17 | | | *Active* | *Passive* |
| 7 | 31 | 15 | | Mean | 24.417 | 17.875 |
| 8 | 15 | 26 | | Variance | 39.819 | 16.201 |
| 9 | 24 | 17 | | Observations | 24 | 24 |
| 10 | 27 | 12 | | df | 23 | 23 |
| 11 | 23 | 25 | | F | 2.458 | |
| 12 | 20 | 21 | | P(F<=f) one-tail | 0.0179 | |
| 13 | 22 | 20 | | F Critical one-tail | 2.312 | |

Figure 7.13: $F$-Test Data and Output

**Solution.** We will use the $F$-test in the **Analysis ToolPak.**

1. Enter the data in columns A and B of a workbook, shown here, with part of the data, in Figure 7.13.

2. From the Menu Bar choose **Data Analysis** – ***F*-Test Two-Sample for Variances** and complete the dialog box of Figure 7.14. Notice that we have

inserted not the specified level of significance $\alpha = 0.05$ in this box but rather half the value, 0.025, to reflect the fact that our test is two-sided while the cells E12:E13 give the $P$-value and the critical value for a one-sided upper-tailed test.

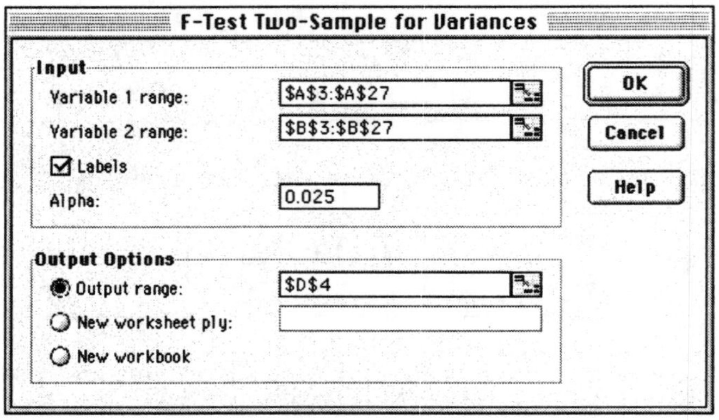

Figure 7.14: $F$-Test Dialog Box

## Caveat

This tool requires that the larger of the two sample variances be in the numerator, so repeat this procedure by reversing the variables if the output shows that the variance of the data in **Variable 1 range** in Figure 7.14 is less than that in **Variable 2 range** (which is not the case here).

## Excel Output

The output appears in cells D4:F13 as shown in Figure 7.13. The computed value of $F$ under $H_0$

$$F = \frac{s_1^2}{s_2^2} = 2.1955$$

appears in E11. (Note that $s_1^2 > s_2^2$ as required.) The critical $F$ value is 2.312 and therefore the data is significant at the 5% level.

We can obtain the $P$-value from cell E12 which gives the one-sided $P$-value, which we need to double and find

$$P - \text{value} = 0.0358$$

## The $F$ Distribution Function

This a good place to record the syntax for the $F$ distribution. Suppose that $F$ is a random variable having an $F$ distribution with degrees of freedom $\nu_1$ for numerator and $\nu_2$ for denominator. Then for any $x > 0$

$$\texttt{FDIST}(x, \nu_1, \nu_2) = P(F > x)$$

while for any $0 < p < 1$ the upper $p$ critical value is obtained from the inverse

$$P\,(F > \texttt{FINV}(p, \nu_1, \nu_2)) = p$$

# Chapter 8

# Inference for Proportions

In this chapter we discuss data representing the counts or proportions of outcomes occurring in different categories. Examples are:

- How common is behavior that puts people at risk of AIDS? The National AIDS Behavioral Survey interviewed a random sample of 2673 adult heterosexuals. Of these, 170 had more than one sexual partner in the past year. Based on these data, what can we say about the percent of all adult heterosexuals who have multiple partners?

- Do preschool programs for poor children make a difference in later life? A study looked at 62 children who were enrolled in a Michigan preschool in the late 1960s and at a control group of 61 similar children who were not enrolled. At 27 years of age, 61% of the preschool group and 80% of the control group had required the help of a social service agency in the previous ten years. Is this significant evidence that preschool for poor children reduces later use of social services?

## 8.1 Inference for a Population Proportion

To estimate the proportion $p$ of some characteristic in a population it is common to take an SRS of size $n$ and count $X$ = the number in the sample possessing the characteristic. For large $n$, the distribution of $X$ is approximately binomial $B(n,p)$ and by the Central Limit Theorem the sample proportion

$$\hat{p} \text{ is approximately } N\left(p, \sqrt{\frac{p(1-p)}{n}}\right)$$

Inference is then based on the procedures for estimating a normal mean discussed in Chapter 6.

## Confidence Intervals

The standard error of $\hat{p}$ is

$$\text{SE} = \sqrt{\frac{\hat{p}(1-\hat{p})}{n}}$$

where we have replaced $p$ with $\hat{p}$ in the expression for the standard deviation of $\hat{p}$. Therefore, a large sample level $C$ confidence interval for $p$ is given as

$$\hat{p} \pm z^*\text{SE}$$

where $z^*$ is the upper $(1-C)/2$ standard normal critical value.

**Example 8.1.** (Exercise 8.19 page 445 in text.) The Gallup Poll asked a sample of 1785 adults, "Did you, yourself, happen to attend church or synagogue in the last 7 days?" Of the respondents, 750 said "Yes." Treat Gallup's sample as an SRS of all American adults.

(a) Give a 99% confidence interval for the proportion of all adults who attended church or synagogue during the week preceding the poll.

(b) Do the results provide good evidence that less than half of the population attended church or synagogue?

(c) How large a sample would be required to obtain a margin of error of 0.01 in a 99% confidence interval for the proportion who attend church or synagogue? (Use the conservative guess of $p^* = 0.50$ and explain why this method is reasonable in this situation.)

| | A | B | C |
|---|---|---|---|
| 1 | | *Confidence Interval for a Population Proportion* | |
| 2 | | | |
| 3 | | Values | Formulas |
| 4 | User Input | | |
| 5 | conf | 0.99 | 0.99 |
| 6 | Summary Statistics | | |
| 7 | n | 1785 | |
| 8 | X | 750 | |
| 9 | Calculations | | |
| 10 | p_hat | 0.4202 | =X/n |
| 11 | SE | 0.0117 | =SQRT(p_hat*(1-p_hat)/n) |
| 12 | z | 2.58 | =NORMSINV(0.5+conf/2) |
| 13 | ME | 0.030 | =z*SE |
| 14 | Confidence Limits | | |
| 15 | lower | 0.390 | =p_hat-ME |
| 16 | upper | 0.450 | =p_hat+ME |
| 17 | | | |
| 18 | Required Sample Size | 16588 | =INT((z/0.01)^2*(0.5)*(1-0.5))+1 |

Figure 8.1: Confidence Interval for a Population Proportion

**Solution.** Figure 8.1 shows the Excel formulas required for the calculation (in column C), together with the corresponding values obtained when these formulas are entered in column B on your workbook. The formulas parallel those for the confidence interval for a normal mean. From cells B15:B16 we find that a 99% confidence interval is (0.390, 0.450). Since the confidence interval lies entirely to the left of the point 0.50 this data provides very good evidence that less than half the population attended services. From the workbook in cell B13 we read that the margin of error is 0.030. To obtain a margin of error 0.01 we would require a sample of size 16588 – see cell B18 – roughly 9 times the size of the sample used because the desired margin of error is 1/3 the margin of error in the sample.

**Note:** We remind the reader that the formulas in Figure 8.1 require **Named Ranges** to refer to the variables by their names, or else the cell references must be used.

## Tests of Significance

A large sample procedure for testing the null hypothesis

$$H_0 : p = p_0$$

uses the test statistic

$$z = \frac{\hat{p} - p_0}{\sqrt{\frac{p_0(1-p_0)}{n}}}$$

For example, if the alternative is two-sided

$$H_a : p \neq p_0$$

then we reject $H_0$ at level $\alpha$ if

$$|z| > z^*$$

where $z^*$ is the upper $\frac{\alpha}{2}$ standard normal critical value. Furthermore, the $P$-value is $2P(Z > |z|)$ where $Z$ is $N(0,1)$.

**Example 8.2.** (Example 8.6 page 438 in text.) A coin that is balanced should come up heads half the time in the long run. The population for coin tossing contains the results of tossing the coin forever. The parameter $p$ is the probability of a head, which is the proportion of all tosses that give a head. The tosses we actually make are an SRS from this population.

The French naturalist Count Buffon (1701 - 1788) tossed a coin 4040 times. He got 2048 heads. The sample proportion of heads is

$$\hat{p} = \frac{2048}{4040} = 0.5069$$

That's a bit more than one-half. Is this evidence that Buffon's coin was not balanced?

| | A | B | C |
|---|---|---|---|
| 1 | | | *Significance Test for a Population Proportion* |
| 2 | | | |
| 3 | User Input | | |
| 4 | p0 | 0.50 | |
| 5 | alpha | 0.01 | |
| 6 | | | |
| 7 | Summary Statistics | | |
| 8 | n | 4040 | |
| 9 | X | 2048 | |
| 10 | Calculations | | |
| 11 | p_hat | 0.5069 | =X/n |
| 12 | SE | 0.0079 | =SQRT(p0*(1-p0)/n) |
| 13 | z | 0.881 | =(p_hat-p0)/SE |
| 14 | Lower Test | | |
| 15 | lower_z | | =NORMSINV(alpha) |
| 16 | Decision | | =IF(z<lower_z,"Reject H0","Do Not Reject H0") |
| 17 | Pvalue | | =NORMSDIST(z) |
| 18 | Upper Test | | |
| 19 | upper_z | | =-NORMSINV(alpha) |
| 20 | Decision | | =IF(z>upper_z,"Reject H0","Do Not Reject H0") |
| 21 | Pvalue | | =1-NORMSDIST(z) |
| 22 | Two-Sided Test | | |
| 23 | two_z | 2.576 | =ABS(NORMSINV(alpha/2)) |
| 24 | Decision | Do Not Reject H0 | =IF(ABS(z)>two_z,"Reject H0","Do Not Reject H0") |
| 25 | Pvalue | 0.378 | =2*(1-NORMSDIST(ABS(z))) |

Figure 8.2: Significance Test for a Population Proportion

**Solution.** To assess whether the data provide evidence that the coin was not balanced, we test

$$H_0 : p = 0.5$$
$$H_a : p \neq 0.5$$

where $p$ is the probability that Buffon's coin lands heads.

Figure 8.2 gives the required formulas in column C for lower, upper, and two-sided tests. Column B contains the data and calculations. Cell B11 gives $\hat{p} = 0.5069$, cell B12 gives the standard error SE $= 0.0079$, and cell B13 gives the calculated value of $z = 0.881$. Our problem is two-sided so only the values in rows 22 – 25 of the template are relevant. The $P$-value is 0.378 and we therefore do not reject $H_0$. These calculations are straightforward and Figure 8.2 merely provides a systematic way of carrying them out. Nothing as elaborate is really needed for calculations which could be done by a hand-calculator.

**Exercise.** Simulate 4040 tosses of a fair coin. Repeat the simulation 100 times and plot the lower and upper end points of the 100 confidence intervals on a graph with the constant line $p = 0.5$. How many intervals contain 0.5?

## 8.2 Comparing Two Proportions

Large sample inference procedures for comparing the proportions $p_1$ and $p_2$ in two populations based on independent SRS of sizes $n_1$ and $n_2$, respectively, are also based on the normal approximation. The natural estimate $\hat{p}_1 - \hat{p}_2$ of the difference in proportions $p_1 - p_2$ is approximately normal with mean $p_1 - p_2$ and standard deviation

$$\sigma = \sqrt{\frac{p_1(1-p_1)}{n_1} + \frac{p_2(1-p_2)}{n_2}}$$

### Confidence Intervals

We must replace the unknown parameters $p_1$ and $p_2$ by their estimates $\hat{p}_1$ and $\hat{p}_2$ to obtain an estimated standard error

$$\mathrm{SE} = \sqrt{\frac{\hat{p}_1(1-\hat{p}_1)}{n_1} + \frac{\hat{p}_2(1-\hat{p}_2)}{n_2}}$$

and an approximate level $C$ confidence interval

$$\hat{p}_1 - \hat{p}_2 \pm z^*\mathrm{SE}$$

where $z^*$ is the upper $(1-C)/2$ standard normal critical value.

> **Example 8.3.** (Examples 8.9 and 8.10 pages 447 - 449 in text.) To study the long-term effects of preschool programs for poor children the High/Scope Educational Research foundation has followed two groups of Michigan children since early childhood. One group of 62 attended preschool as 3- and 4-year-olds. This is a sample from Population 2: poor children who attend preschool. A control group of 61 children from the same area and similar backgrounds represents Population 1: poor children with no preschool. One response variable of interest is the need for social services as adults. In the past 10 years, 38 of the preschool sample and 49 of the control sample have needed social services (mainly welfare). If $p_1$ and $p_2$ represent the proportions of Population 1 (resp. Population 2) needing social services as adults, then the difference $p_1 - p_2$ is a measure of the effect of preschool in reducing the proportion of children who need social services later as adults. Find a 95% confidence interval for $p_1 - p_2$.

**Solution.** Table 8.1 conveniently summarizes the data. The corresponding Excel workbook is in Figure 8.3. Results of the calculations are given in column B with the corresponding formulas in the adjacent column C. Based on this data set we are 95% confident that the percent needing social services is somewhere between 3.3% and 34.7% lower among people who attended preschool.

Table 8.1: How Much Does Preschool Help?

| Population | Description | $n$ | $X$ | $\hat{p} = X/n$ |
|---|---|---|---|---|
| 1 | Control | 61 | 49 | 0.803 |
| 2 | Preschool | 62 | 38 | 0.613 |
| | Total | 123 | 87 | 0.707 |

| | A | B | C | D | E |
|---|---|---|---|---|---|
| 1 | *Confidence Interval for the Difference in Proportions* | | | | |
| 2 | | | | | |
| 3 | Summary Statistics and User Input | | | | |
| 4 | Population | n | X | p_hat=X/n | |
| 5 | Control | 61 | 49 | 0.8033 | |
| 6 | Preschool | 62 | 38 | 0.6129 | |
| 7 | | | | | |
| 8 | conf | 0.95 | | | |
| 9 | SE | 0.08011 | =SQRT((D5*(1-D5)/B5)+(D6*(1-D6)/B6))) | | |
| 10 | z | 1.960 | =NORMSINV(0.5+conf/2) | | |
| 11 | ME | 0.157 | =z*SE | | |
| 12 | | | | | |
| 13 | Confidence Limits | | | | |
| 14 | lower | 0.033 | =(D5-D6)-ME | | |
| 15 | upper | 0.347 | =(D5-D6)+ME | | |
| 16 | | | | | |
| 17 | | | | | |

Figure 8.3: Confidence Interval – Difference in Proportions

## Tests of Significance

The null hypothesis

$$H_0 : p_1 = p_2$$

is tested using the statistic

$$z = \frac{\hat{p}_1 - \hat{p}_2}{\text{SE}}$$

where SE is the standard error based on the pooled estimate

$$\hat{p} = \frac{x_1 + x_2}{n_1 + n_2}$$

of the common value $p \equiv p_1 = p_2$ of the population proportions. Here $x_1$ and $x_2$ are the number of counts possessing the characteristic being counted, in sample 1, sample 2, respectively, and

$$\text{SE} = \sqrt{\hat{p}(1 - \hat{p})\left(\frac{1}{n_1} + \frac{1}{n_2}\right)}$$

The decision rules based on $z$ are then analogous to those in Section 8.1. For example, if the alternative hypothesis is

$$H_a : p_1 < p_2$$

then we reject at level $\alpha$ if

$$z < -z^*$$

where $z^*$ is the upper $\alpha$ standard normal critical value. Furthermore, the $P$-value is $P(Z < z)$ where $Z$ is $N(0,1)$.

**Example 8.4.** (Examples 8.11 and 8.12 in text.) High levels of cholesterol in the blood are associated with higher risk of heart attacks. Will using a drug to lower blood cholesterol reduce heart attacks? The Helinski Heart Study looked at this question. Middle-aged men were assigned at random to one of two treatments: 2051 men took the drug gemfibrozil to reduce their cholesterol levels, and a control group of 2030 men took a placebo. During the next five years, 56 men in the gemfibrozil group and 84 men in the placebo group had heart attacks.

| | A | B | C | D | E |
|---|---|---|---|---|---|
| 1 | | | *Significance Test for the Difference in Proportions* | | |
| 2 | | | | | |
| 3 | | Summary Statistics and User Input | | | |
| 4 | Group | n | X | p_hat | |
| 5 | Gemfibrozil | 2051 | 56 | 0.0273 | |
| 6 | Placebo | 2030 | 84 | 0.0414 | |
| 7 | Total | 4081 | 140 | | |
| 8 | null | 0 | Calculations | | |
| 9 | alpha | 0.05 | pooled_p | 0.0343 | =(C5+C6)/(B5+B6) |
| 10 | | | SE | 0.0057 | =SQRT(pooled_p*(1-pooled_p)*(1/B5 + 1/B6)) |
| 11 | | | z | -2.470 | =((D5-D6)-null)/SE |
| 12 | Lower Test | | | | |
| 13 | lower_z | -1.645 | =NORMSINV(alpha) | | |
| 14 | Decision | Reject H0 | =IF(z<lower_z,"Reject H0","Do Not Reject H0") | | |
| 15 | Pvalue | 0.0068 | =NORMSDIST(z) | | |
| 16 | Upper Test | | | | |
| 17 | upper_z | | =-NORMSINV(alpha) | | |
| 18 | Decision | | =IF(z>upper_z,"Reject H0","Do Not Reject H0") | | |
| 19 | Pvalue | | =1-NORMSDIST(z) | | |
| 20 | Two-Sided Test | | | | |
| 21 | two_z | | =ABS(NORMSINV(alpha/2)) | | |
| 22 | Decision | | =IF(ABS(z)>two_z,"Reject H0","Do Not Reject H0") | | |
| 23 | Pvalue | | =2*(1-NORMSDIST(ABS(z))) | | |

Figure 8.4: Significance Test – Difference in Proportions

**Solution.** Let $p_1$ and $p_2$ be the proportions in hypothetical gemfibrozil (resp. placebo) populations of men who would suffer heart attacks during a corresponding five-year period as the men in the sample.

Since we are trying to "prove" that gemfibrozil reduces the incidence of heart attacks, the appropriate significance test should be

$$H_0 : p_1 = p_2$$
$$H_a : p_1 < p_2$$

Figure 8.4 is a template which provides all the formulas required. Although this is a lower-tailed test, we have provided the formulas for upper-tailed and two-tailed tests. From cells B5:B6 we read

$$\hat{p}_1 = \frac{56}{2051} = 0.0273$$

$$\hat{p}_2 = \frac{84}{2030} = 0.0414$$

and cell B9 gives the pooled sample proportion

$$\hat{p} = \frac{140}{4081} = 0.0343$$

The $z$-test statistic in cell D11 is $-2.470$ and the $P$-value in cell B15 is 0.0068. This is considered statistically significant so there is strong evidence that gemfibrozil reduced the rate of heart attacks.

# Chapter 9

# Inference for Two-Way Tables

Example 8.3 was a comparison of two populations of children (control, preschool) with respect to one response variable "need of social services as an adult." The response variable had two values, "needed" or "not needed." A test was carried out of the null hypothesis

$$H_0 : p_1 = p_2 \tag{9.1}$$

where $p_1$ and $p_2$ are the proportions requiring social services as adults in the respective two populations.

We wish to generalize this test to the situation in which there are more than two populations of interest or where the response variable can take more than two values. Table 9.1 presents the data of Example 8.3 in a slightly different (though equivalent) way than was presented in Table 8.1.

Table 9.1: How Much Does Preschool Help? $2 \times 2$ Table

| | Social Services | | |
|---|---|---|---|
| Population | Needed | Not Needed | Total |
| Control | 49 | 12 | 61 |
| Preschool | 24 | 38 | 62 |
| Total | 73 | 50 | 123 |

This presentation of the data shows that there are two variables of interest, one which might be considered as an explanatory variable (control versus preschool) and the other as the response variable. The significance test of the null hypothesis (9.1) judges whether there is a relationship between the two variables. If $p_1 = p_2$ then there is no relationship.

A table such as Table 9.1 showing data collected on two categorical variables having $r$ rows and $c$ columns of values for each of the two variables is called an $r \times c$ table. In this chapter we discuss a technique based on the chi-square distribution for deciding if there is a relationship between two categorical variables.

**Example 9.1.** (Example 9.1 page 470 in text.) Cocaine addicts need the drug to feel pleasure. Perhaps giving them a medication that fights depression will help them stay off cocaine. A three-year study compared an antidepressant called desipramine with lithium (a standard treatment for cocaine addiction) and a placebo. The subjects were 72 chronic users of cocaine who wanted to break their drug habit. Twenty-four of the subjects were randomly assigned to each treatment. Table 9.2 gives the counts of the subjects who relapsed or avoided replase into cocaine use during the study. What is the null hypothesis of interest? Find the table of expected counts if the null hypothesis is true and compare with the observed counts. Calculate the proportion of subjects in each of the three treatment groups who had no relapse.

Table 9.2: Treating Cocaine Addiction

| | Response | | |
|---|---|---|---|
| Treatment | No Relapse | Relapse | Total |
| Desipramine | 14 | 10 | 24 |
| Lithium | 6 | 18 | 24 |
| Placebo | 4 | 20 | 24 |
| Totals | 24 | 48 | 72 |

In this example the explanatory variable "Treatment" has three levels, in contrast to the corresponding variable "Population" in Table 9.1. The response variable has two levels "No Relapse" and "Relapse" displayed across the top of Table 9.2, just as in Table 9.1. The objective is to determine whether the responses are statistically the same for all treatment levels. Another possible way of expressing this objective is to say that we are asking whether the cross classifications into the two variables are independent of each other. Which interpretation is appropriate depends on how the data are collected, essentially whether with fixed or random marginal totals, but the methodology is the same in either case.

## 9.1 Two-Way Tables

Each combination of row and column levels in Table 9.2 defines a cell. The data in the three rows labelled "Desipramine," "Lithium," and "Placebo" represent the results of independent samples from respective populations of individuals given these treatments.

Suppose that

$$\begin{aligned} \mathbf{p}_1 &= (p_{11}, p_{12}) \\ \mathbf{p}_2 &= (p_{21}, p_{22}) \\ \mathbf{p}_3 &= (p_{31}, p_{32}) \end{aligned}$$

represent the population proportions among the three treatment groups with responses, "No Relapse" and "Relapse," respectively. Thus, for instance, $p_{12} = \frac{10}{48} = 0.2083$ = the proportion of a hypothetical Desipramine treated population who would relapse into cocaine use.

The null hypothesis of interest is

$$H_0 : \mathbf{p}_1 = \mathbf{p}_2 = \mathbf{p}_3 \tag{9.2}$$

meaning equality of the three vectors of proportions, thereby generalizing (9.1). With $r$ rows we would be testing

$$H_0 : \mathbf{p}_1 = \mathbf{p}_2 = \cdots = \mathbf{p}_r$$

for corresponding population proportion vectors $\mathbf{p}_i$. Also, note that since there are only two possible responses in this example, then since

$$\begin{aligned} p_{11} &= 1 - p_{12} \\ p_{21} &= 1 - p_{22} \\ p_{31} &= 1 - p_{32} \end{aligned}$$

we are really only testing equality of three scalars

$$H_0 : p_{11} = p_{21} = p_{31}$$

However, with a more general $c \geq 3$ columns (response levels) we would be testing equality of true vectors of proportions as in (9.2).

Some preparation is needed before we can use the Excel function `CHIINV` which returns the $P$-value required to carry out the test in Example 9.1. We need to first create a table of expected cell counts.

## Expected Cell Counts

Consider column 2 in Table 9.2, "Relapse," in which there are 10 counts under Desipramine, 18 under Lithium, and 20 under Placebo for a total column sum of 48. Under the null hypothesis (9.2), the three treatment levels have the same effect, so we may judge their overall effect by pooling. We therefore get a pooled estimate

$$\frac{10 + 18 + 20}{72} = \frac{48}{72} = 0.6667$$

for the common value $p_{12} = p_{22} = p_{32}$. Similarly, the pooled estimate is

$$\frac{14+6+4}{72} = \frac{24}{72} = 0.3333$$

for the common value $p_{11} = p_{21} = p_{31}$.

The row total for Desipramine is 24. Therefore, the number of counts in cell "Desipramine × Relapse" is a binomial random variable on 24 trials with "success probability" $\frac{48}{72}$. Under $H_0$ we estimate this with the usual binomial mean

$$24 \times \frac{48}{72} = \frac{24 \times 48}{72} = 16$$

leading to the useful mnemonic

$$\text{expected count} = \frac{\text{row total} \times \text{column total}}{\text{table total}}$$

Expected counts need to be computed for all cells. These calculations need to be hand-coded in the Excel workbook and we now show how to do this. Copy the observed counts to an Excel workbook. This is shown in Figure 9.1 where the data appear in block A3:C7 (including labels).

1. For the row totals type the formula = SUM(B5:C5) in cell D5 and press Enter.

2. Select cell D5, then click the fill handle and fill down column B to cell B7.

3. Enter the formula = SUM(B5:B7) in cell B8 and fill across row 8 to cell D8 to produce the row totals and the overall table total.

Next we produce the table of expected counts. Again refer to Figure 9.1.

| | A | B | C | D | E | F | G | H | I |
|---|---|---|---|---|---|---|---|---|---|
| 1 | | | | | | *Expected Table* | | | |
| 2 | | | | | | | | | |
| 3 | | Observed Counts | | | | | Formulas | | |
| 4 | | No Relapse | Relapse | Row Total | | | No Relapse | Relapse | Row Total |
| 5 | Desipramine | 14 | 10 | 24 | | Desipramine | =B$8*$D5/$D$8 | =C$8*$D5/$D$8 | =SUM(B12:C12) |
| 6 | Lithium | 6 | 18 | 24 | | Lithium | =B$8*$D6/$D$8 | =C$8*$D6/$D$8 | =SUM(B13:C13) |
| 7 | Placebo | 4 | 20 | 24 | | Placebo | =B$8*$D7/$D$8 | =C$8*$D7/$D$8 | =SUM(B14:C14) |
| 8 | Column Total | 24 | 48 | 72 | | Column Total | =SUM(B12:B14) | =SUM(C12:C14) | =SUM(D12:D14) |
| 9 | | | | | | | | | |
| 10 | | Expected Counts | | | | | | | |
| 11 | | No Relapse | Relapse | Row Total | | Chi-Square Test | | | |
| 12 | Desipramine | 8 | 16 | 24 | | P-value | 0.005 | | |
| 13 | Lithium | 8 | 16 | 24 | | | =CHITEST(B5:C7,B12:C14) | | |
| 14 | Placebo | 8 | 16 | 24 | | | | | |
| 15 | Column Total | 24 | 48 | 72 | | | | | |

Figure 9.1: Expected Cell Counts and $P$-value

1. Copy the entire observed table block A3:D8 to another location, say A10:D15 and change the label "Observed Counts" to "Expected Counts".

2. Select cell B12 and enter the formula "= B$8*$D5/$D$8" which involves both absolute and relative cell reference. Fill the formula to the other cells in the block B12:B14 by first filling from cell B12 to C12 and then from block B12:C12 to B14:C14. This fills the table with expected counts. We require an absolute reference $8 since the column totals are always in row 8. We require an absolute reference $D because the row totals are always in column D. We don't want these to change when the formula is filled to all the cells in the expected table, so absolute references are mandated as shown.

3. Finally obtain the marginal sums as before.

Figure 9.1 shows both the values as well as the formulas which are behind the cells in the table of expected counts.

## 9.2 The Chi-Square Test

The statistic

$$X^2 = \sum \frac{(\text{observed count} - \text{expected count})^2}{\text{expected count}}$$

where the sum is taken over counts in all the cells was introduced by Karl Pearson in 1900 to measure how well the model $H_0$ fits the data. Under $H_0$ it has a sampling distribution which is approximated by a one-parameter family of distributions known as chi-square and denoted by the symbol $\chi^2$. The parameter is called the degrees of freedom $\nu$, and for an $r \times c$ contingency table it is known that $\nu = (r-1)(c-1)$. If $H_0$ is true then $X^2$ should be "small," while if $H_0$ is false then $X^2$ should be "large." This leads to the criterion:

$$\text{chi-square test:} \quad \text{Reject } H_0 \text{ if } X^2 > \chi^2_\alpha$$

where $\chi^2_\alpha$ is the upper critical $\alpha$ value of a chi-square distribution on $(r-1)(c-1)$ degrees of freedom.

### The $P$-Value

The Excel function `CHITEST` calculates the $P$-value associated with the Pearson $X^2$ statistic. The syntax is

= `CHITEST`(actual_range, expected_range)

where "actual_range" refers to the observed table of counts B5:C7 and where "expected_range" refers to the expected table of counts B12:C14. Refer to block F11:G13 of Figure 9.1 where we have given the formula and its value. The $P$-value is 0.005, so we reject $H_0$ and conclude that there are differences in the success rates for Desipramine, Lithium, and Placebo.

## Computing the Value of $X^2$

Although not required for the function CHITEST, it is instructive to do the calculation of $X^2$ and then carry out the test based on value of $X^2$. We have shown this in block F3:H12 on the right half of Figure 9.2. The required formulas are displayed in Figure 9.3 and we present the details.

| | A | B | C | D | E | F | G | H |
|---|---|---|---|---|---|---|---|---|
| 1 | | | *Computing Chi-Square* | | | | | |
| 2 | | | | | | | | |
| 3 | | Relapse | | | | Chi-Square Cell Values | | |
| 4 | | No Relapse | Relapse | Row Total | | | NoRelapse | Relapse |
| 5 | Desipramine | 14 | 10 | 24 | | Desipramine | 4.50 | 2.25 |
| 6 | Lithium | 6 | 18 | 24 | | Lithium | 0.50 | 0.25 |
| 7 | Placebo | 4 | 20 | 24 | | Placebo | 2.00 | 1.00 |
| 8 | Column Total | 24 | 48 | 72 | | | | |
| 9 | | | | | | Chi-Square | 10.500 | |
| 10 | | Expected Counts | | | | Critical 5% | 5.991 | |
| 11 | | No Relapse | Relapse | Row Total | | | | |
| 12 | Desipramine | 8 | 16 | 24 | | Decision: | Reject H0 | |
| 13 | Lithium | 8 | 16 | 24 | | | | |
| 14 | Placebo | 8 | 16 | 24 | | | | |
| 15 | Column Total | 24 | 48 | 72 | | | | |

Figure 9.2: Computing the Value of $X^2$

| | F | G | H |
|---|---|---|---|
| 1 | | | |
| 2 | | | |
| 3 | | Chi-Square Cell Values | |
| 4 | | No Relapse | Relapse |
| 5 | Desipramine | =(B5-B12)^2/B12 | =(C5-C12)^2/C12 |
| 6 | Lithium | =(B6-B13)^2/B13 | =(C6-C13)^2/C13 |
| 7 | Placebo | =(B7-B14)^2/B14 | =(C7-C14)^2/C14 |
| 8 | | | |
| 9 | Chi-Square | =SUM(G5:H7) | |
| 10 | Critical 5% | =CHIINV(0.05,2) | |
| 11 | | | |
| 12 | Decision: | =IF(G9>G10, "Reject H0", "Do Not Reject H0") | |

Figure 9.3: Computing the Value of $X^2$ – Formulas

1. Copy the block A4:C7 to a convenient location, shown here copied to cells F4:H7. Change the label "Response" to "Chi-Square Cell Values."

2. The equation for $X^2$ is

$$X^2 = \sum \frac{(\text{observed count} - \text{expected count})^2}{\text{expected count}}$$

which we translate into Excel by the formula "= (B5-B12)^ 2/B12" entered in cell G5 and then filled to the block G5:H7. See Figure 9.3.

3. Sum all six cell entries by entering = `SUM(G5:H7)` in cell G9. We get the value

$$X^2 = 10.500$$

The critical 5% $\chi^2$ value is given by = `CHIINV(0.05,2)` in cell G10 and is

$$\chi^2_{.05} = 5.991$$

We therefore reject $H_0$, which is the same conclusion drawn using $P$-value.

You may arrange for the decision to appear on your workbook using the formula = `IF`($G9 > G10$, "Reject H0", "Do Not Reject H0") which we have entered in cell G12. For convenience we have shown all required formulas in Figure 9.3 which is a portion of Figure 9.2 showing the formulas behind the cells.

# Chapter 10

# One-Way Analysis of Variance: Comparing Several Means

The two-sample $t$ test of Chapter 7 compared the means of two populations. Analysis of Variance (ANOVA) is a technique for comparing the means of more than two populations and is a direct generalization of the two-sample $t$ test. In particular, the $F$ statistic in ANOVA when there are two populations is precisely the square of the Student $t$, and the ANOVA $F$ test is then identical to the Student two-sample $t$ procedure.

As the name suggests, ANOVA consists of separating the variability in a data set into two components and judging whether a fit that assumes equal population means is substantially better than a fit in which all means are assumed the same. This is achieved by comparing the residual variation following the model fit in both cases by a ratio called an $F$, which may be viewed as a signal-to-noise ratio.

## 10.1 The Analysis of Variance $F$ Test

Suppose there are $I$ normal populations labelled $1 \le i \le I$ and that independent random samples $\{x_{ij} : 1 \le j \le n_i\}$ of size $n_i \ge 1$, $1 \le i \le I$ are taken from each. The ANOVA model is

$$x_{ij} = \mu_i + \varepsilon_{ij} \qquad 1 \le i \le I,\ 1 \le j \le n_i$$

where $\{\varepsilon_{ij}\}$ are independent $N(0, \sigma)$ random variables. In words, the populations have means $\mu_i$ and a common standard deviation $\sigma$, and we express this as

$$DATA = FIT + RESIDUAL$$

The one-way ANOVA significance test is

$$H_0 : \mu_1 = \mu_2 = \cdots = \mu_I$$
$$H_a : \text{ not all of the } \mu_i \text{ are equal.}$$

Under $H_0$ we estimate the common value $\mu$ with the overall mean

$$\bar{x} = \sum_{i=1}^{I} \sum_{j=1}^{n_i} x_{ij}$$

What is left over $x_{ij} - \bar{x}$ is called the residual under $H_0$, and then the total residual variation in the data is given by

$$\text{SST} = \sum_{i=1}^{I} \sum_{j=1}^{n_i} (x_{ij} - \bar{x})^2$$

called the **total sum of squares.**

If we do not assume $H_0$ is true, then we should estimate each individual $\mu_i$ by the corresponding sample mean $\bar{x}_i$. The remainder $x_{ij} - \bar{x}_i$ is then the residual under an unrestricted model, and the total residual variation in the data is given by

$$\text{SSE} = \sum_{i=1}^{I} \sum_{j=1}^{n_i} (x_{ij} - \bar{x}_i)^2$$

called the **error** (or within groups) **sum of squares.** Intuitively, a better fit always occurs with the unrestricted model. We can show with a little algebra that the difference SSG = SST − SSE is positive and can also be expressed as

$$\text{SSG} = \sum_{i=1}^{I} \sum_{j=1}^{n_i} (\bar{x}_i - \bar{x})^2$$

called the **between groups sum of squares.** Remarkably it turns out that

$$\text{SST} = \text{SSG} + \text{SSE}$$

which is the key to the partition of the variation.

The magnitude of SSG measures the improvement in the fit as measured by the residual sum of squares. In order to reject $H_0$, the improvement must be significantly beyond what might be expected due to chance, and one is led to consider the ratio $\frac{\text{SSG}}{\text{SSE}}$ . Define the mean squares

$$\text{MSG} = \frac{\text{SSG}}{I-1} \qquad \text{MSE} = \frac{\text{SSE}}{N-I}$$

where $N = \sum_{i=1}^{I} n_i$, and form the ratio

$$F = \frac{\text{MSG}}{\text{MSE}}$$

known to have an $F$ distribution with $I-1$ degrees of freedom for the numerator and $N-I$ degrees of freedom for the denominator (denoted by $F(I-1, N-I)$).

The decision rule is

$$\text{Reject } H_0 \text{ at level } \alpha \text{ if } F > F^*$$

where $F^*$ is the upper $\alpha$ critical value of an $F(I-1, N-I)$ distribution, that is, $F^*$ satisfies

$$P\left(F(I-1, N-I) > F^*\right) = \alpha.$$

We observe that MSE can also be expressed as

$$\text{SSE} = \sum_{i=1}^{I}(n_i - 1)s_i^2$$

where $\{s_i\}$ are the sample variances and, consequently, MSE is a pooled sample variance

$$s_p^2 \equiv \text{MSE} = \frac{\sum_{i=1}^{I}(n_i-1)s_i^2}{\sum_{i=1}^{I}(n_i-1)}$$

and therefore an unbiased estimate of $\sigma^2$.

Finally, we define the ANOVA coefficient of determination

$$r^2 = \frac{\text{SSG}}{\text{SST}}$$

as the fraction of the total variance "explained by model $H_0$".

## Carrying out an ANOVA

We illustrate the implementation of the preceding discussion with the following worked exercise.

**Example 10.1.** (Exercise 10.17 page 526 in text.) How do nematodes (microscope worms) affect plant growth? A botanist prepares 16 identical planting pots and then introduces different numbers of nematodes into the pots. He transplants a tomato seedling into each plot. Table 10.1 shows the data on the increase in height of seedlings (in centimeters) 16 days after planting (data provided by Matthew Moore).

(a) Make a table of means and standard deviations for the four different nematode samples, and plot the data and the means.
(b) State $H_0$ and $H_a$ for the ANOVA test for these data, and explain in words what ANOVA tests in this setting.
(c) Using Excel, carry out the ANOVA. What are the $F$ statistic and its $P$-value? State the values of $s_p$ and $r^2$. Report your overall conclusions about the effect of nematodes on plant growth.

Table 10.1: Nematodes and Tomato Plants

| Nematodes | Seedling growth (cm) | | | |
|---|---|---|---|---|
| 0 | 10.8 | 9.1 | 13.5 | 9.2 |
| 1000 | 11.1 | 11.1 | 8.2 | 11.3 |
| 5000 | 5.4 | 4.6 | 7.4 | 5.0 |
| 10000 | 5.8 | 5.3 | 3.2 | 7.5 |

## Plotting the Data and the Sample Means

The first step in ANOVA is usually exploratory, where the data and means are plotted. Such a plot will help to visually confirm or dispel the assumption of equal variances and to indicate possible outliers or skewness in the data that might call into question use of this technique.

This is easily carried out in Excel using the **ChartWizard**, which produces side-by-side displays of the samples $\{x_{ij}\}$, their sample means $\{\bar{x}_i\}$, and the overall mean $\bar{x}$. These provide a good preliminary display of ANOVA data.

### Means and Standard Deviations

The means and standard deviations are evaluated with the Excel functions `AVERAGE` and `STDEV`. We refer to Figure 10.1 throughout.

1. Enter the data and labels in B3:E7 of a workbook (Figure 10.1).
2. Enter the label "mean" in A8 and then the formula = `AVERAGE`(B4:B7) in B8. The value 10.65 appears, which is the sample mean of the "0" observations. Select cell B8 and, using the fill handle, drag to E8, filling the cells with the means for the other nematode numbers.
3. Enter the label "stdev" in A9 and the formula = `STDEV`(B4:B7) in B9. Then select B9 and drag the fill handle to E9. Now the standard deviations of all the samples appear.

### Plotting

The workbook needs to be prepared for the **ChartWizard** by relocating the data, coding the samples, relocating the sample means, and entering the overall mean. Refer to Figure 10.1 and complete columns G, H, I, and J, as indicated.

Since the sample means have already been calculated in B8:E8, an efficient way to relocate them to I19:I22 is to select B8:E8, choose **Edit-Copy** from the Menu Bar, and then select cell I19 and choose **Edit-Paste Special** from the Menu Bar. Complete the **Paste Special** dialog box as in Figure 10.2. The radio button for values needs to be selected because the contents of B8:E8 are formulas, not

| | A | B | C | D | E | F | G | H | I | J |
|---|---|---|---|---|---|---|---|---|---|---|
| 1 | | | | *One-Way Analysis of Variance* | | | | | | |
| 2 | | | Nematodes | | | | Nematodes | Growth | | |
| 3 | | 0 | 1000 | 5000 | 10000 | | 1 | 10.8 | Means | Overall |
| 4 | | 10.8 | 11.1 | 5.4 | 5.8 | | 1 | 9.1 | | |
| 5 | | 9.1 | 11.1 | 4.6 | 5.3 | | 1 | 13.5 | | |
| 6 | | 13.5 | 8.2 | 7.4 | 3.2 | | 1 | 9.2 | | |
| 7 | | 9.2 | 11.3 | 5.0 | 7.5 | | 2 | 11.1 | | |
| 8 | mean | 10.650 | 10.425 | 5.600 | 5.450 | | 2 | 11.1 | | |
| 9 | stdev | 2.053 | 1.486 | 1.244 | 1.771 | | 2 | 8.2 | | |
| 10 | | | | | | | 2 | 11.3 | | |
| 11 | | | | | | | 3 | 5.4 | | |
| 12 | | | | | | | 3 | 4.6 | | |
| 13 | | | | | | | 3 | 7.4 | | |
| 14 | | | | | | | 3 | 5.0 | | |
| 15 | | | | | | | 4 | 5.8 | | |
| 16 | | | | | | | 4 | 5.3 | | |
| 17 | | | | | | | 4 | 3.2 | | |
| 18 | | | | | | | 4 | 7.5 | | |
| 19 | | | | | | | 1 | | 10.650 | 8.031 |
| 20 | | | | | | | 2 | | 10.425 | 8.031 |
| 21 | | | | | | | 3 | | 5.600 | 8.031 |
| 22 | | | | | | | 4 | | 5.450 | 8.031 |

Figure 10.1: Preparing the Data for the One-Way ANOVA Tool

values, and a straight **Edit-Copy** will change the relative cell references. The check box **Transpose** merely converts a row selection into a column. Finally, the common value 8.031 in J19:J22 is the overall mean obtained from the Excel formula = `AVERAGE`(B4:E7).

**Paste Special**

**Paste**

- ○ All
- ○ Formulas
- ● Values
- ○ Formats
- ○ Comments
- ○ Validation
- ○ All except borders

**Operation**

- ● None
- ○ Add
- ○ Subtract
- ○ Multiply
- ○ Divide

☐ Skip blanks ☐ Transpose

Paste Link | Cancel | OK

Figure 10.2: Paste Special

1. Begin as always by clicking the **ChartWizard** button.

2. **Users of Excel 5**

   - In Step 1 of 5 enter G2:J22 for the range. This will produce a simultaneous scatterplot with all three variables – data, sample means, overall mean – plotted on the $y$ axis (with different markers) against the nematode sample code (column G) on the $x$ axis.

- In Step 2 select **XY (Scatter)**.
- In Step 3 select Format **1**.
- Complete Step 4 as follows: Data Series in **Columns**, First "1" Column for X Data, and First "1" Row for Legend Text
- Complete Step 5 as follows: Add a Legend? **Yes**, Chart Title "Nematodes and Growth: Data and Means," Axis Title Category (X) "Number of Nematodes," and Axis Title Value (Y) "Seedling Growth (cm)." Click Finish

**Users of Excel 97/98**

- In Step 1 select **XY (Scatter)** for Chart Type and the upper left Chart sub-type **Scatter**.
- In Step 2 under under the **Data Range** tab enter G2:J22 (which will produce a simultaneous scatterplot with all three variables, data, sample means, and overall mean, plotted on the y axis (with different markers) against the nematode sample code (column G) on the x axis, and select the Series radio button for **Columns**.
- In Step 3 under the **Titles** tab, type "Number of Nematodes" as the Chart Title and leave blank both **Value (X) Axis** and **Value (Y) Axis**. Under the **Axes** tab, both check boxes should be selected. Under the **Gridlines** tab clear all check boxes. Under the **Legend** tab check the **Show legend** and locate it to the Right. Finally, under the **Data Labels** tab select the radio button **None**.
- In Step 4 embed the graph in the current workbook by selecting the radio button **As object in**. Click Finish.

A scatterplot like the one shown in Figure 10.3 appears with distinct markers representing the individual observations, the sample means, and the overall mean. Editing will enhance the usefulness of this plot.

### Editing the Plot

We edit Figure 10.3 to more clearly emphasize the observations, the sample means, and the overall mean. The result resembles side-by-side boxplots.

1. Activate the chart for editing.
2. Click one of the markers representing a sample mean (the entire series will then appear highlighted). From the Menu Bar choose **Format – Selected Data Series....** In the dialog box under the **Patterns** tab select radio buttons **Automatic** for **Line** and **None** for **Marker**. Click OK. The markers for sample means are now replaced by a connecting polygonal line.

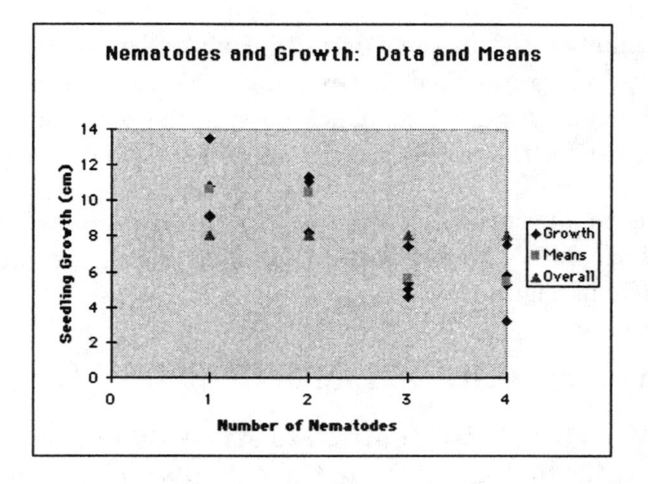


Figure 10.3: Default Plot – Data and Means

3. Repeat this step after selecting a marker for the overall mean. The four markers are replaced by a horizontal line.

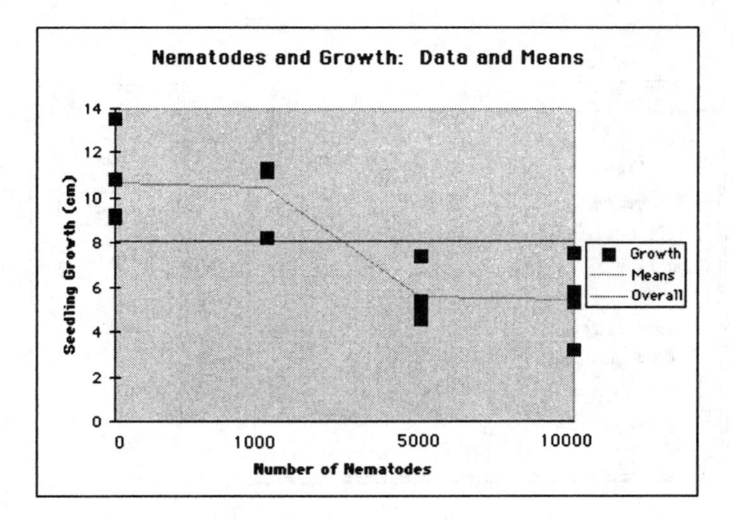


Figure 10.4: Enhanced Plot – Data and Means

4. Click a marker for the observations and change the Style of the marker to a square.

5. Select the horizontal axis and from the Menu Bar choose **Format-Selected Axis. . . .** In the dialog box click the **Scale** tab and type "1" for **Minimum.** Then click the **Patterns** tab and click **None** for **Tick-Mark Labels**, and click **None** for **Major** and **Minor Tick Mark Type**. Click OK.

6. Type the number "0" in the **Formula Bar** and press enter. The number "0" appears in a greyshaded text rectangle. With the cursor, drag it to the location shown in Figure 10.4. Click outside the text rectangle to de-select it. Now repeat for the other three labels "1000," "5000," and "10000."

The result of this enhancement is Figure 10.4, which immediately shows the dominant features of the data set much more aggressively than the numerical counterpart in cells A3:E9 of Figure 10.1.

## Using the One-Way ANOVA Tool

1. From the Menu Bar choose **Tools-Data Analysis** and select **Anova:Single Factor** from the tools listed in the **Data Analysis** dialog box. Click OK to display the Anova:Single Factor dialog box (Figure 10.5).

2. Type the cell references B3:E7 (or point and drag on the workbook) for **Input range**. Select the radio button Grouped By: **Columns**. Check the box **Labels in first row** and use "0.05" for Alpha. Enter K2 for **Output range**. Click OK.

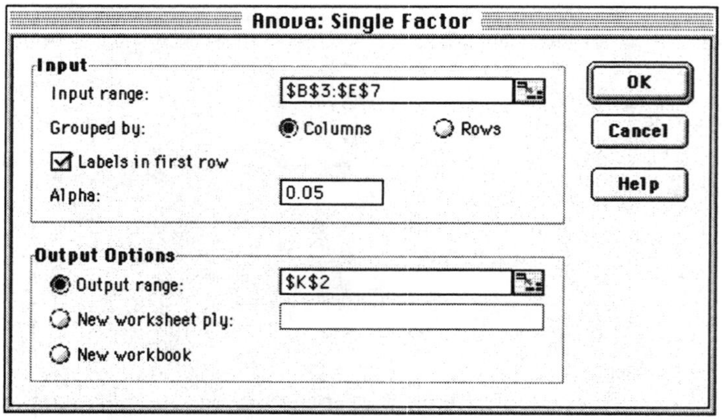

Anova: Single Factor
Input
Input range: $B$3:$E$7
Grouped by: Columns Rows
Labels in first row
Alpha: 0.05
Output Options
Output range: $K$2
New worksheet ply:
New workbook
OK
Cancel
Help

Figure 10.5: Single Factor ANOVA Dialog Box

## Excel Output – ANOVA Table

The output from the **Anova:Single Factor** tool appears in Figure 10.6 from which we can read off all the variables described earlier in this chapter. The first half of the output provides summary statistics for the four samples. Had we not desired a plot, we would have read the sample means and standard deviations directly from this output.

| | K | L | M | N | O | P | Q |
|---|---|---|---|---|---|---|---|
| 1 | | | | | | | |
| 2 | Anova: Single Factor | | | | | | |
| 3 | | | | | | | |
| 4 | SUMMARY | | | | | | |
| 5 | *Groups* | *Count* | *Sum* | *Average* | *Variance* | | |
| 6 | 0 | 4 | 42.60 | 10.650 | 4.217 | | |
| 7 | 1000 | 4 | 41.70 | 10.425 | 2.209 | | |
| 8 | 5000 | 4 | 22.40 | 5.600 | 1.547 | | |
| 9 | 10000 | 4 | 21.80 | 5.450 | 3.137 | | |
| 10 | | | | | | | |
| 11 | | | | | | | |
| 12 | ANOVA | | | | | | |
| 13 | *Source of Variation* | *SS* | *df* | *MS* | *F* | *P-value* | *F crit* |
| 14 | Between Groups | 100.647 | 3 | 33.549 | 12.08 | 0.00062 | 3.490 |
| 15 | Within Groups | 33.328 | 12 | 2.777 | | | |
| 16 | | | | | | | |
| 17 | Total | 133.9744 | 15 | | | | |

Figure 10.6: Single Factor ANOVA Output

**Source of Variation**

Between Groups sum of squares refers to

$$\text{SSG} = 100.647$$

Within Groups (error) sum of squares is

$$\text{SSE} = 33.328$$

The Total sum of squares is

$$\text{SST} = 133.9744$$

After dividing by the respective degrees of freedom (*df*), we arrive at the mean squares (*MS*)

$$\begin{aligned} \text{MSG} &= \frac{100.647}{3} = 33.549 \\ \text{MSE} &= \frac{33.328}{20} = 2.777 \end{aligned}$$

and the calculated $F$ statistic is

$$F = \frac{\text{MSG}}{\text{MSE}} = \frac{33.549}{2.777} = 12.08$$

The 5% critical $F$ value ($F$ *crit*) is

$$F^* = 3.490$$

The $P$-value is 0.00062.

We conclude that for the significance test

$$H_0 : \mu_1 = \mu_2 = \mu_3 = \mu_4$$
$$H_a : \text{not all of the } \mu_i \text{ are equal}$$

we reject $H_0$ at the 5% level (in fact at any reasonable level), in view of the small $P$-value.

We observe that the pooled estimate of variance is

$$s_p^2 = \text{ MSE } = 2.777$$

which we note is also (because of the equal sample sizes) the average of the individual sample variances.

Finally, although the coefficient of determination is not provided, we can do the arithmetic to derive

$$r^2 = \frac{\text{SSG}}{\text{SST}} = \frac{100.647}{133.9744} = 0.751.$$

# Chapter 11

# Inference for Regression

We pointed out in Chapter 2 that there are three procedures in Excel for regression analysis. These are complementary for the most part. We have already used the **Insert Trendline** command to fit and draw the least-squares regression line. In this chapter we derive additional information about the regression model with the **Regression** tool in the ToolPak. We also introduce some relevant Excel functions.

## 11.1 Inference About the Model

The general regression model for $n$ pairs of observation $(x_i, y_i)$ is

$$DATA = FIT + RESIDUAL$$

which is expressed mathematically as

$$y_i = \alpha + \beta x_i + \varepsilon_i.$$

The function

$$\mu_y = \alpha + \beta x$$

is called the population (true) regression curve of $y$ on $x$, here taken to be linear. It represents the mean response of $y$ as a function of $x$. The quantities $\{\varepsilon_i\}$ are assumed to be independent normally distributed random variables with a common mean 0 and common standard deviation $\sigma$. Therefore there are three unknown parameters – $\alpha, \beta$, and $\sigma$.

The parameters $\alpha$ and $\beta$ are estimated by the method of least-squares (described in Chapter 2) by the values $a$ and $b$, respectively, where

$$\begin{aligned} a &= \bar{y} - b\bar{x} \\ b &= r\,\frac{s_y}{s_x}\,. \end{aligned}$$

The least-squares regression line estimates the population regression line, which is given by the equation

$$\hat{y} = a + bx.$$

Algebra shows that $a$ and $b$ are unbiased estimators of $\alpha$ and $\beta$, respectively. They are random variables having sampling distributions:

(i) $a$ is normal with

$$\begin{aligned} \text{mean} &= \mu_a = \alpha \\ \text{standard deviation} &= \sigma_a = \sigma\sqrt{\frac{1}{n} + \frac{\bar{x}^2}{\sum_{i=1}^{n}(x_i - \bar{x})^2}} \end{aligned}$$

(ii) $b$ is normal with

$$\begin{aligned} \text{mean} &= \mu_b = \beta \\ \text{standard deviation} &= \sigma_b = \frac{\sigma}{\sqrt{\sum_{i=1}^{n}(x_i - \bar{x})^2}} \end{aligned}$$

(iii) In a manner entirely analogous to the case of estimating the mean of a data set, we require an estimate of $\sigma^2$ for inference purposes. This estimate is provided by $s^2$,

$$s^2 = \frac{\sum_{i=1}^{n} e_i^2}{n-2}$$

where $e_i = y_i - \hat{y}_i$ is called the residual at $x_i$ and is an estimate of the sampling error $\{\varepsilon_i\}$. The denominator $(n-2)$ arises because

$$\frac{(n-2)s^2}{\sigma^2} \quad \text{is chi-squared on } (n-2) \; df.$$

## The Regression Tool

**Example 11.1.** (Examples 11.1 - 11.5 beginning on page 530 in text.) Infants who cry easily may be more easily stimulated than others and this may be a sign of a higher IQ. Child development researchers explored the relationship between the crying of infants four to ten days old and their later IQ scores. A snap of a rubber band on the sole of the foot caused the infants to cry. The researchers recorded the crying and measured its intensity by the number of peaks in the most active 20 seconds. They later measured the children's IQ at age three years using the Stanford-Binet IQ test. Table 11.1 contains data on 38 infants. Analyse the data using the **Regression** tool as follows:

(a) Plot the data and confirm that a straight line fit is appropriate.
(b) Fit the least-squares regression line to the data.
(c) Find the standard error.
(d) Give 95% confidence intervals for the parameters $\beta$ and $\alpha$.
(e) Test the null hypothesis $H_0 : \beta = 0$.
(f) Compute $r^2$. Examine the residuals for any deviation from a straight line.

Table 11.1: Infants Crying and IQ Scores

| Crying | IQ | Crying | IQ | Crying | IQ | Crying | IQ |
|---|---|---|---|---|---|---|---|
| 10 | 87 | 20 | 90 | 17 | 94 | 12 | 94 |
| 12 | 97 | 16 | 100 | 19 | 103 | 12 | 103 |
| 9 | 103 | 23 | 103 | 13 | 104 | 14 | 106 |
| 16 | 106 | 27 | 108 | 18 | 109 | 10 | 109 |
| 18 | 109 | 15 | 112 | 18 | 112 | 23 | 113 |
| 15 | 114 | 21 | 114 | 16 | 118 | 9 | 119 |
| 12 | 119 | 12 | 120 | 19 | 120 | 16 | 124 |
| 20 | 132 | 15 | 133 | 22 | 135 | 31 | 135 |
| 16 | 136 | 17 | 141 | 30 | 155 | 22 | 157 |
| 33 | 159 | 13 | 162 | | | | |

## Using the Regression Tool

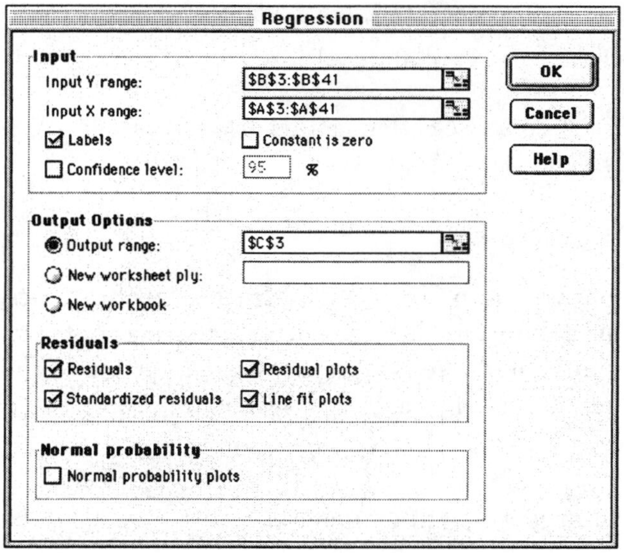

Figure 11.1: Regression Tool Dialog Box

1. Enter the 38 data pairs (including labels) from Table 11.1 into columns of a workbook, with the independent variable $x$ to the left of the dependent variable $y$, say in columns A3:A41 and B3:B41. (Excel requires this kind of data in contiguous regions.) Later, refer to Figure 11.2, which shows the **Regression** tool output, as well as columns A and B.

2. From the Menu Bar choose **Tools – Data Analysis** and select **Regression** from the tools listed in the **Data Analysis** dialog box. Click OK to display the **Regression** dialog box (Figure 11.1).

3. Type the cell references B3:B41 in Figure 11.1 (or point and drag over the data in column B) for **Input Y range**. Do the same with respect to the first column for **Input X range**. Check the box **Labels**, leave **Constant is zero** clear because we are not forcing the line through the origin, check the box **Confidence level** and insert "95." Under **Output options** select the radio button **Output range** and enter C3 to locate the upper left corner where the output will appear in the same workbook.

   Check the following boxes:

   Residuals. To obtain predicted or fitted values $\hat{y}$ and their residuals.

   Residual Plots. To generate a scatterplot of the residuals against their $x$ values.

   Standardized Residuals. To obtain residuals divided by their standard error (useful to identify outliers).

   Line Fit Plots. To generate a scatterplot of $y$ against $x$.

   Do not check Normal Probability Plots.

**Note:** In making the above selections either use the tab key to move from option to option or use the mouse. Click OK.

## Excel Output

The output is separated into six regions: Regression Statistics, ANOVA table, statistics about the regression line parameters, residual output, scatterplot with fitted line, and residual plot. We have reproduced a portion of the output, without the graphs or residuals, in Figure 11.2. We now interpret the output in each of these regions.

### Parameter Estimates and Inference

Rows 19 and 20 of the output provide statistics for the regression line parameters. From cell D19 we read $a = 91.27$, and from cell D20 we read $b = 1.493$. The regression line therefore is

$$\hat{y} = 91.27 + 1.493x.$$

## Significance Tests

| | A | B | C | D | E | F | G | H | I |
|---|---|---|---|---|---|---|---|---|---|
| 1 | | | | *Regression Example. Infants' Crying and IQ Scores* | | | | | |
| 2 | | | | | | | | | |
| 3 | Crying | IQ | SUMMARY OUTPUT | | | | | | |
| 4 | 10 | 87 | | | | | | | |
| 5 | 12 | 97 | *Regression Statistics* | | | | | | |
| 6 | 9 | 103 | Multiple R | 0.4550 | | | | | |
| 7 | 16 | 106 | R Square | 0.2070 | | | | | |
| 8 | 18 | 109 | Adjusted R Squ | 0.1850 | | | | | |
| 9 | 15 | 114 | Standard Error | 17.4987 | | | | | |
| 10 | 12 | 119 | Observations | 38 | | | | | |
| 11 | 20 | 132 | | | | | | | |
| 12 | 16 | 136 | ANOVA | | | | | | |
| 13 | 33 | 159 | | *df* | *SS* | *MS* | *F* | *Significance F* | |
| 14 | 20 | 90 | Regression | 1 | 2877.480 | 2877.480 | 9.397 | 0.004 | |
| 15 | 16 | 100 | Residual | 36 | 11023.389 | 306.205 | | | |
| 16 | 23 | 103 | Total | 37 | 13900.868 | | | | |
| 17 | 27 | 108 | | | | | | | |
| 18 | 15 | 112 | | *Coefficients* | *Standard Error* | *t Stat* | *P-value* | *Lower 95%* | *Upper 95%* |
| 19 | 21 | 114 | Intercept | 91.27 | 8.934 | 10.216 | 0.000 | 73.15 | 109.39 |
| 20 | 12 | 120 | Crying | 1.493 | 0.487 | 3.065 | 0.004 | 0.505 | 2.481 |
| 21 | 15 | 133 | | | | | | | |

Figure 11.2: Data and Regression Tool Output

The test

$$H_0 : \beta = 0 \qquad vs \qquad H_a : \beta \neq 0$$

is useful in assessing whether a simple model of a straight line through the origin provides an equally good fit. The appropriate test statistic

$$t = \frac{b}{\mathrm{SE}_b}$$

where

$$\mathrm{SE}_b = \frac{s}{\sqrt{\sum_{i=1}^{n}(x_i - \bar{x})^2}}$$

is the standard error of $b$, which is obtained from the standard deviation of $b$ by replacing the unknown $\sigma$ with $s$. From E20 in Figure 11.2 we read $\mathrm{SE}_b = 0.487$, and from F20 we have the $t$ statistic, denoted as $t$ stat,

$$t = \frac{b}{\mathrm{SE}_b} = \frac{1.493}{0.487} = 3.065.$$

The corresponding two sided $P$-value appears in cell G20,

$$P\text{-value} = 0.004$$

Similarly for testing the null hypothesis

$$H_0 : \alpha = 0 \qquad vs \qquad H_a : \alpha \neq 0$$

the appropriate test statistic is

$$t = \frac{a}{\mathrm{SE}_a}$$

where

$$\mathrm{SE}_a = s\sqrt{\frac{1}{n} + \frac{\bar{x}^2}{\sum_{i=1}^{n}(x_i - \bar{x})^2}}$$

estimates the true standard deviation $\sigma_a$ of $a$. From D19:G19 we read the relevant quantities

$$t = \frac{a}{\mathrm{SE}_a} = \frac{91.27}{8.934} = 10.216$$

and two-sided

$$P\text{-value} = 0.000$$

Significance tests for other null values can be carried out using the templates developed in Chapter 7 that required only the summary statistics. The main change is to use $n-2$ instead of $n-1$ for the degrees of freedom.

### Confidence Intervals

We used the default of 95% for the confidence level when we completed the **Regression** tool dialog box. The lower and upper 95% confidence limits appear in H19:I20 in Figure 11.2. Thus, 95% confidence intervals are

$$\begin{array}{ll} \text{for } \beta & (0.505, 2.481) \\ \text{for } \alpha & (73.15, 109.39) \end{array}$$

### Scatterplot

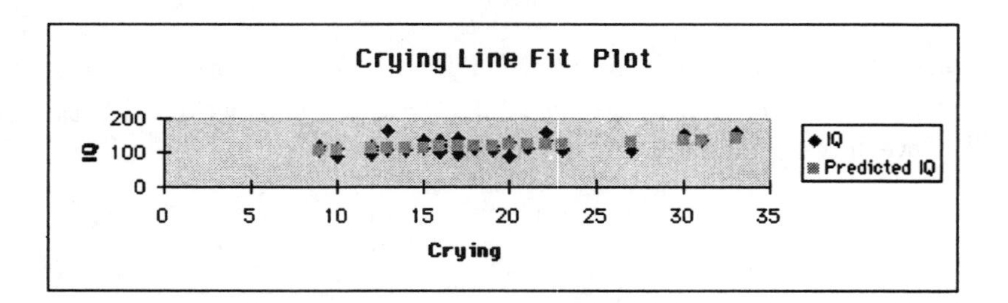


Figure 11.3: Default Scatterplot

Figure 11.3 shows one of the scatterplots produced – IQ against Crying. By default Excel uses markers even for the predicted values and the scales need to be changed to produce a more useful graph.

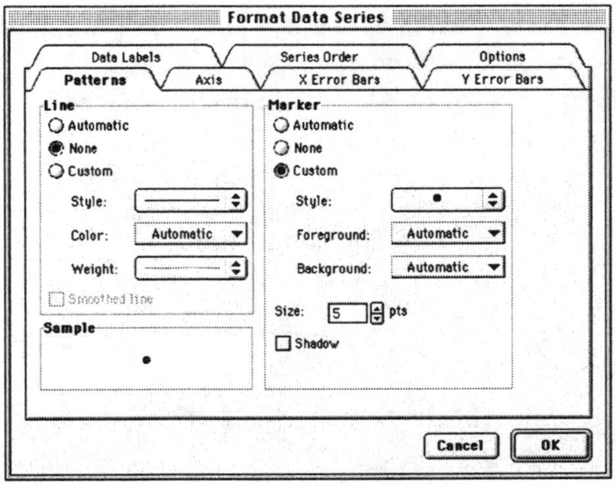

Figure 11.4: Formatting Markers

### Changing Markers

We have modified the Excel scatterplot by enlarging it to make it more readable, changing the scale of the X and Y axes, replacing the diamond-shaped data plotting markers with circles, and replacing the square-shaped predicted values with a line.

1. To resize the **Chart Area** activate the Chart and drag one or more of the handles to the desired size.

2. To resize the **Plot Area** click within the plot area and drag one or more handles to the desired size.

3. To change the scale on the X axis activate the Chart and double-click the X axis. (Equivalently, click the X axis once and choose **Format – Selected Axis...** from the Menu Bar.) Then under the **Scale** tab change **Minimum** to 5 and **Maximum** to 35. Similarly edit the Y-axis and under the **Scale** tab set **Minimum** to 80 and **Maximum** to 180.

4. To change the data markers first activate the Chart and select one of the data points. From the Menu Bar choose **Format – Selected Data Series...**, click the **Patterns** tab, and select a **Custom Marker** (Figure 11.4).

### Changing Predicted Markers to Line

The other important enhancement to the markers is to change the predicted ones in the default scatterplot into a line. This is a useful enhancement even in other contexts and we separate its description here. The steps are as follows.

1. Activate the Chart and select one of the predicted markers. Choose **Format-Selected Data Series...** from the Menu Bar .

2. In the **Format Data Series** dialog box click the **Patterns** tab. Select radio button **Custom** for **Line** and then pick a Color. Finally select **None** for **Marker**. The markers are disappear and are replaced by the regression line (Figure 11.5).

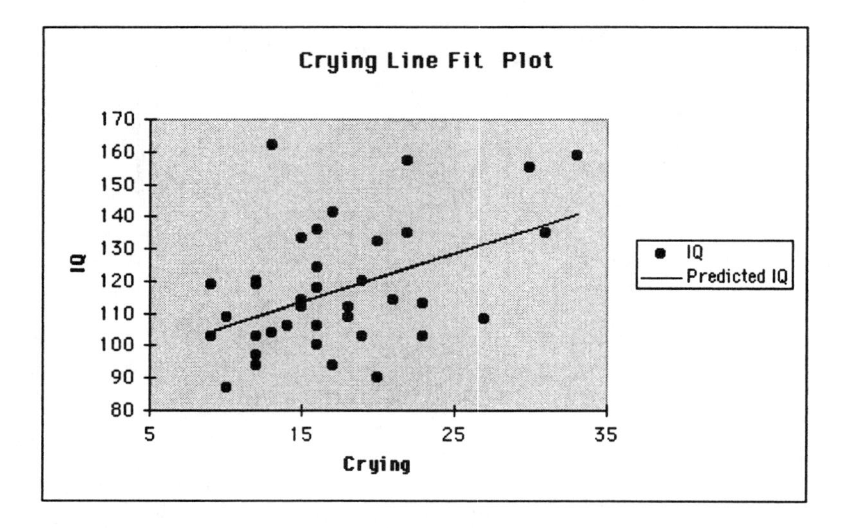


Figure 11.5: Regression Line and Enhanced Scatterplot

## Residual Plot

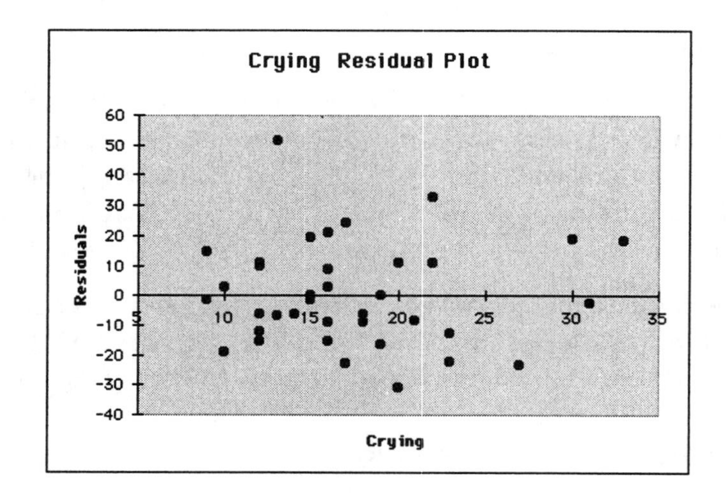


Figure 11.6: Residual Plot

The enhanced residual plot appears in Figure 11.6. The residuals show a random scatter about the X axis indicating that the straight line fit is appropriate.

### Residual Output

| | C | D | E | F |
|---|---|---|---|---|
| 24 | RESIDUAL OUTPUT | | | |
| 25 | | | | |
| 26 | *Observation* | *Predicted IQ* | *Residuals* | *Standard Residuals* |
| 27 | 1 | 106.197265 | -19.19726463 | -1.112199386 |
| 28 | 2 | 109.183058 | -12.18305783 | -0.705829174 |
| 29 | 3 | 104.704368 | -1.704368032 | -0.098743082 |
| 30 | 4 | 115.154644 | -9.154644218 | -0.530377107 |
| 31 | 5 | 118.140437 | -9.140437415 | -0.529554031 |
| 32 | 6 | 113.661748 | 0.33825238 | 0.019596755 |
| 33 | 7 | 109.183058 | 9.816942174 | 0.568747541 |
| 34 | 8 | 121.126231 | 10.87376939 | 0.62997148 |
| 35 | 9 | 115.154644 | 20.84535578 | 1.20768205 |
| 36 | 10 | 140.533886 | 18.46611361 | 1.069839928 |
| 37 | 11 | 121.126231 | -31.12623061 | -1.803307671 |
| 38 | 12 | 115.154644 | -15.15464422 | -0.877988938 |
| 39 | 13 | 125.60492 | -22.60492041 | -1.309622963 |
| 40 | 14 | 131.576507 | -23.5765068 | -1.365912117 |
| 41 | 15 | 113.661748 | -1.66174762 | -0.096273856 |
| 42 | 16 | 122.619127 | -8.619127209 | -0.499351766 |
| 43 | 17 | 109.183058 | 10.81694217 | 0.626682846 |
| 44 | 18 | 113.661748 | 19.33825238 | 1.120367554 |

Figure 11.7: Residual Output

Figure 11.7 shows a portion of the residual output with all predicted values, residuals and standardized residuals. With this at hand, other diagnostic scatterplots can be quickly obtained, in addition to the default output – for example, residuals versus $x$ variable.

### ANOVA Table

The ANOVA approach uses an $F$-test to determine whether a substantially better fit is obtained by the regression model than by a model with $\beta = 0$. The ANOVA output breaks the observed total variation in the data

$$\text{SST} = \sum_{i=1}^{n} (y_i - \bar{y})^2$$

into two components, residual or error sum of squares

$$\text{SSE} = \sum_{i=1}^{n} (y_i - \hat{y}_i)^2$$

and a model sum of squares

$$\text{SSM} = \sum_{i=1}^{n} (\hat{y}_i - \bar{y})^2$$

connected by the identity

$$\text{SST} = \text{SSE} + \text{SSM}.$$

The test criterion is based on how much smaller the residual sum of squares is under each fit and is based on the $F$-ratio

$$F = \frac{\text{MSM}}{\text{MSE}}$$

where $\text{MSM} = \frac{\text{SSM}}{1}$ is the mean square for the model fit, while $\text{MSE} = \frac{\text{SSE}}{n-2}$ is the mean square for error. Under the null hypothesis that $\beta = 0$ this ratio has an $F$ distribution with 1 degree of freedom in the numerator and $n-2$ degrees of freedom in the denominator.

| | C | D | E | F | G | H |
|---|---|---|---|---|---|---|
| 12 | ANOVA | | | | | |
| 13 | | *df* | *SS* | *MS* | *F* | *Significance F* |
| 14 | Regression | 1 | 2877.480 | 2877.480 | 9.397 | 0.004 |
| 15 | Residual | 36 | 11023.389 | 306.205 | | |
| 16 | Total | 37 | 13900.868 | | | |

Figure 11.8: ANOVA Output

All the above calculations appear in rows 13–16 in Figure 11.8. This is an advanced topic. The only point we add here is that the $F$ test is identical to the earlier two-sided test for $H_0 : \beta = 0$ *vs* $H_a : \beta \neq 0$, and the $F$ statistic 9.397 in cell G14 is always the square of the $t$-stat 3.065 appearing in cell F20 in Figure 11.2 in this context.

### Regression Statistics

This is the last component of the output from Figure 11.2 that we discuss briefly and isolate in Figure 11.9, which gives, for instance, the coefficient of determination $r^2 = 0.2070$ and the standard error $s = 17.4987$, in addition to more advanced statistics such as the Adjusted $R$ Square used in multiple regression.

| | C | D |
|---|---|---|
| 5 | *Regression Statistics* | |
| 6 | Multiple R | 0.4550 |
| 7 | R Square | 0.2070 |
| 8 | Adjusted R Square | 0.1850 |
| 9 | Standard Error | 17.4987 |
| 10 | Observations | 38 |

Figure 11.9: Regression Statistics Output

## 11.2 Inference About Predictions

In the preceding section we drew inferences on the parameters $\alpha$ and $\beta$ in the regression line $\mu_y = \alpha + \beta x$. Here we examine the mean response and the prediction of a single outcome, both at a specified value $x^*$ of the explanatory variable.

- To estimate the mean response we use a confidence interval for the parameter $\mu_y$ based on the point estimate

$$\hat{\mu_y} \equiv \hat{y} = a + bx$$

A level $C$ confidence interval for $\mu_y$ is given by

$$\hat{y} \pm t^* SE_{\hat{\mu}} \tag{11.1}$$

where the standard error is given by

$$\mathrm{SE}_{\hat{\mu}} = s\sqrt{\frac{1}{n} + \frac{(x^* - \bar{x})^2}{\sum_{i=1}^{n}(x_i - \bar{x})^2}}$$

- To predict a single observation at $x^*$ a level $C$ prediction interval given by

$$\hat{y} \pm t^* \mathrm{SE}_{\hat{y}} \tag{11.2}$$

where the appropriate standard error is now given by

$$\mathrm{SE}_{\hat{y}} = s\sqrt{1 + \frac{1}{n} + \frac{(x^* - \bar{x})^2}{\sum_{i=1}^{n}(x_i - \bar{x})^2}}$$

In each case $t^*$ is the upper $(1 - C)/2$ critical value of the Student $t$-distribution with $n - 2$ degrees of freedom.

Unfortunately Excel's Regression tool does not provide either of these intervals and (11.1) and (11.2) need to be constructed using Excel functions and appropriate cell references.

There are several ways to proceed. The Regression tool gives, inter alia, $a, b$, and $s$ from which $\hat{y}$ and $\mathrm{SE}_{\hat{\mu}}$ can be determined. We illustrate with Example 11.2 below.

An alternative to the Regression tool uses Excel regression functions. This is the third method mentioned in Chapter 2 for regression analysis and is taken up in detail at the end of this section.

**Example 11.2.** (Exercise 11.2 page 535 in text.) Table 11.2 in the text (reproduced in Figure 11.10 contains data on the natural gas consumption of the Sanchez houshold for 16 months. Gas consumption is higher in cold weather. The table gives the average amount of natural gas consumed each day during the month, in hundreds of cubic feet, and the average number of heating degree-days each day during the month. After these data were collected, the Sanchez family installed solar panels.

(a) In the month of January after the solar panels were installed there were 40 degree-days per day. How much gas do you predict the Sanchez household would have used per day without the solar panels?

(b) Give a 95% prediction interval for the amount of gas that would have been used this January without solar panels.

(c) Give a 95% confidence interval for the mean gas consumption per day in all months with 40 degree-days per day.

| | A | B | C | D | E | K |
|---|---|---|---|---|---|---|
| 1 | | | *Confidence and Prediction Intervals* | | | |
| 2 | | | | | | |
| 3 | Degree-days | Gas | SUMMARY OUTPUT | | | |
| 4 | 24 | 6.3 | | | | |
| 5 | 51 | 10.9 | *Regression Statistics* | | | |
| 6 | 43 | 8.9 | Multiple R | 0.9953 | | |
| 7 | 33 | 7.5 | R Square | 0.9906 | | |
| 8 | 26 | 5.3 | Adjusted R Square | 0.9899 | | |
| 9 | 13 | 4.0 | Standard Error | 0.3389 | | |
| 10 | 4 | 1.7 | Observations | 16 | | |
| 11 | 0 | 1.2 | | | | |
| 12 | 0 | 1.2 | ANOVA | | | |
| 13 | 1 | 1.2 | | *df* | *SS* | |
| 14 | 6 | 2.1 | Regression | 1 | 168.58116 | |
| 15 | 12 | 3.1 | Residual | 14 | 1.60821 | |
| 16 | 30 | 6.4 | Total | 15 | 170.18938 | |
| 17 | 32 | 7.2 | | | | |
| 18 | 52 | 11.0 | | *Coefficients* | *Standard Error* | |
| 19 | 30 | 6.9 | Intercept | 1.08921 | 0.13891 | |
| 20 | | | Degree-days | 0.18900 | 0.00493 | |
| 21 | | | | | | |
| 22 | Prediction Interval for a Future Observation | | | | | |
| 23 | Fit | 8.6492 | =D19+D20*40 | | | |
| 24 | StDev Fit | 0.3601 | =D9*SQRT(1+1/16 + (40-AVERAGE(A4:A19))^2/(16*VARP(A4:A19))) | | | |
| 25 | PI lower limit | 7.8768 | =B23-TINV(0.05,14)*B24 | | | |
| 26 | PI upper limit | 9.4215 | =B23+TINV(0.05,14)*B24 | | | |
| 27 | | | | | | |
| 28 | Confidence Interval for Mean Response | | | | | |
| 29 | Fit | 8.6492 | =D19+D20*40 | | | |
| 30 | StDev Fit | 0.1216 | =D9*SQRT(1/16 + (40-AVERAGE(A4:A19))^2/(16*VARP(A4:A19))) | | | |
| 31 | CI lower limit | 8.3883 | =B29-TINV(0.05,14)*B30 | | | |
| 32 | CI upper limit | 8.9100 | =B28+TINV(0.05,14)*B30 | | | |

Figure 11.10: Confidence and Prediction Intervals

**Solution.**

1. Enter the 16 pairs of observations from Table 11.2 of your text into columns A, B of a workbook (Figure 11.10).

2. Following the instructions given in Example 11.1 invoke the **Regression** tool, completing the dialog box without checking the boxes for residuals, and output the results beginning in cell C3. Figure 11.10 shows the relevant portion of the output. In this workbook the value of $a$ is in cell D19, $b$ is in D20, and $s$ is in D9.

3. In a blank cell, B23 in Figure 11.10, type the formula

$$= D19 + D20 * 40$$

giving a predicted value $\hat{y} = 8.6492$. In cell B24 enter the formula

$$= \texttt{D9} * \texttt{SQRT}(1 + 1/16 + (40 - \texttt{AVERAGE}(A4 : A19))\hat{\ } 2/(16 * \texttt{VARP}(A4 : A19)))$$

giving $\text{SE}_{\hat{y}} = 0.1216$. In B25 enter the formula

$$= B23 - \texttt{TINV}(0.05, 14) * B24$$

and in B26 enter the formula

$$= B23 + \texttt{TINV}(0.05, 14) * B24$$

(For convenience we have shown these formulas on the workbook.) You can read the lower and upper 95% prediction interval (7.8768, 9.4215) in B25:B26.

4. Entirely analgous steps are used to derive a confidence interval for the mean gas consumption per day in all months with 40 degree-days per day, the only difference being in the formula for the standard error $\text{SE}_{\mu}$. Again, all formulas are shown in Figure 11.10. The required 95% confidence interval is (8.3883, 8.9100).

## Regression Functions

A third method for regression analysis uses Excel functions. Suppose we wish to

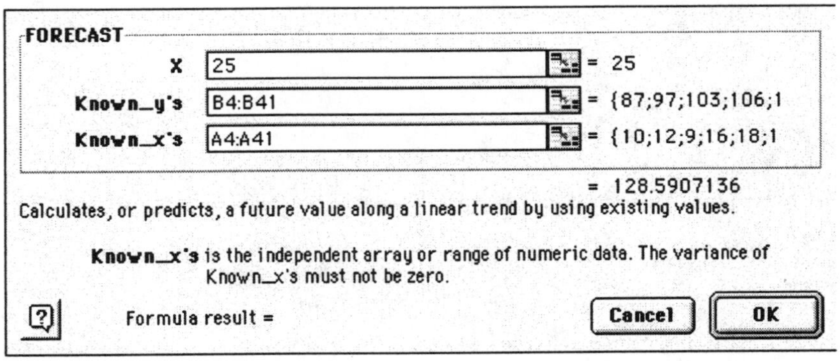

Figure 11.11: The FORECAST Dialog Box

predict IQ at age three years when the count of crying peaks is 25. Without having to derive the regression line and calculating

$$\hat{y} = 91.27 + 1.493(25) = 128.59$$

we could use the **FORECAST** function whose syntax is

$$\texttt{FORECAST}(x, \text{Known\_}y\text{'s}, \text{Known\_}x\text{'s}).$$

Here "Known_$x$'s", refers to the set $\{x_i\}$ while "Known_$y$'s" refers to the $\{y_i\}$. The function can be called from the **Function Wizard** or the **Formula Palette** in which case, you should select Statistical for Function category, FORECAST for Function name, and insert the required parameters in the dialog box (Figure 11.11).

A more general method for obtaining predicted values is TREND with syntax

$$\texttt{TREND}(\text{Known\_}y\text{'s}, \text{Known\_}x\text{'s}, \text{New\_}x\text{'s}, \text{Const}).$$

The parameter "New_$x$'s" is the range of $x$ values for which predictions are desired. The parameter "Const" determines whether the regression line is forced through the origin. Use the value 1 for general $\beta$.

**Example 11.3.** (Example 11.2 continued.) Predict the gas consumption when the average number of heating degree-days during the month are 10, 20, 30, and 40.

D3 = {=TREND(B3:B18,A3:A18,C3:C6)}

| | A | B | C | D | E |
|---|---|---|---|---|---|
| 1 | | | *Using the TREND Function* | | |
| 2 | Degree-days | Gas | x | predicted | |
| 3 | 24 | 6.3 | 10 | 2.979 | |
| 4 | 51 | 10.9 | 20 | 4.869 | |
| 5 | 43 | 8.9 | 30 | 6.759 | |
| 6 | 33 | 7.5 | 40 | 8.649 | |
| 7 | 26 | 5.3 | | | |
| 8 | 13 | 4.0 | | | |
| 9 | 4 | 1.7 | | | |
| 10 | 0 | 1.2 | | | |
| 11 | 0 | 1.2 | | | |
| 12 | 1 | 1.2 | | | |
| 13 | 6 | 2.1 | | | |
| 14 | 12 | 3.1 | | | |
| 15 | 30 | 6.4 | | | |
| 16 | 32 | 7.2 | | | |
| 17 | 52 | 11.0 | | | |
| 18 | 30 | 6.9 | | | |

Figure 11.12: The TREND Function

**Solution.**

1. Select a region of cells D3:D6 for the output.

2. Click the **Formula Palette**, select the TREND function and enter the values B4:B19, A4:A19, and C4:C7 into the argument fields "Known_$y$'s," "Known_$x$'s," and "New_$x$'s," respectively. Leave the "Const" field blank. Click OK.

3. Click the mouse pointer in the **Formula Bar**, hold down the **Shift and Control** keys (either **Macintosh or Windows**), and press enter to **Array-Enter** the formula. The formula will appear surrounded by braces in the Formula Bar indicating it has been array-entered and the predicted values {2.979, 4.869. 6.759, 8.649} will appear in the output range (Figure 11.12).

The last regression function described here is `LINEST` which returns the esti-

C4 = {=LINEST(B4:B19,A4:A19,,TRUE)}

| | A | B | C | D | E |
|---|---|---|---|---|---|
| 1 | | | *Using the LINEST Function* | | |
| 2 | | | | | |
| 3 | Degree-days | Gas | | | |
| 4 | 24 | 6.3 | 0.18900 | 1.08921 | |
| 5 | 51 | 10.9 | 0.00493 | 0.13891 | |
| 6 | 43 | 8.9 | 0.9906 | 0.3389 | |
| 7 | 33 | 7.5 | 1467.551 | 14 | |
| 8 | 26 | 5.3 | 168.581 | 1.60821 | |
| 9 | 13 | 4.0 | | | |
| 10 | 4 | 1.7 | b | a | |
| 11 | 0 | 1.2 | SE_b | SE_a | |
| 12 | 0 | 1.2 | R Square | s | |
| 13 | 1 | 1.2 | F | df | |
| 14 | 6 | 2.1 | SSM | SSE | |
| 15 | 12 | 3.1 | | | |
| 16 | 30 | 6.4 | | | |
| 17 | 32 | 7.2 | | | |
| 18 | 52 | 11.0 | | | |
| 19 | 30 | 6.9 | | | |

Figure 11.13: The `LINEST` Function

mated least-squares line coefficients, their standard errors, $r^2$, $s$, the computed $F$ statistic with its degrees of freedom, and the regression and error sums of squares. The syntax is

$$\texttt{LINEST}(x, \text{Known\_}y\text{'s}, \text{known\_}x\text{'s}, \text{Const}, \text{Stats}).$$

If the field "Stats" is set to false then only the regression coefficients are output while true will produce an entire block of output. The default is false.

**Example 11.4.** (Example 11.2 continued.) Use `LINEST` to perform a regression analysis of gas consumption against the average number of heating degree-days for the data in Figure 11.10.

**Solution.**

1. Select a block of cells C4:E8 with two columns and five rows. (For multiple regression you require a block in which the number of colunns equals the number of independent variables plus one.)

2. Click the **Formula Palette**, select the `LINEST` function and enter the values B4:B19 and A4:A19 into the argument fields "Known_$y$'s" and "Known_$x$'s", respectively. Leave the "Const" field blank and type true in the "Stats" field. Click OK.

3. The formula `= LINEST`(B4:B19,A4:A19) will be visible in the Formula Bar but only the value 0.18900 in cell C4 will appear in your sheet. With the block C4:E8 still selected click the mouse pointer in the **Formula Bar**, as you did for the `TREND` function, hold down the **Shift and Control** keys, and press enter to **Array-Enter** the formula. The formula will appear surrounded by braces in the Formula Bar indicating it has been array-entered and the output will appear in C4:C8 (Figure 11.13)

In cells C10:D14 on Figure 11.13 we have described the contents of the values in C4:E8. These may be compared with the output from the **Regression** tool for this data set, shown in Figure 11.10.

For instance cell D8 Figure 11.13 contains the value 1.60821. This represents the error sum of squares SSE, as seen from the correspond description in D14, which is identical to the output in cell E15 of the **Regression** tool output in Figure 11.10

**Note.** Both `TREND` and `LINEST` can be used with multiple regression.

# Chapter 12

# Nonparametric Tests

Nonparametric procedures are based on the ranks of observations and replace assumptions of normality with less stringent ones such as symmetry and continuity of distributions. Excel does not provide nonparametric tests. Nonetheless, these tests may readily be developed.

## 12.1 The Wilcoxon Rank Sum Test

The Wilcoxon rank sum test is a procedure for comparing independent samples from two populations. Suppose

$$(x_{11}, x_{12}, \ldots, x_{1n_1}), (x_{21}, x_{22}, \ldots, x_{2n_2})$$

are the samples. Generally the $x_1$-observations might be the control group, while the $x_2$-observations are the treatment group.

We wish to test whether both samples can be assumed to arise from identical populations or whether the populations are shifted by a constant. The precise assumption is that if $F_i$ represents the cumulative distribution function of the the $x_i$-observations, $1 \leq i \leq 2$, then

$$F_2(t) = F_1(t + \Delta) \qquad \text{for all real } t$$

where $\Delta$ is a constant representing an unknown shift in the two distributions. If $\Delta$ is positive then the second sample would contain systematically higher values than the first sample.

The null hypothesis is thus

$$H_0 : \mu_1 - \mu_2 = 0$$

where $\mu_1$ and $\mu_2$ are the two population means (or medians).

It is also possible to test a non-zero difference

$$H_0 : \mu_1 - \mu_2 = \Delta_0$$

by subtracting $\Delta_0$ from each $x_1$-observation and applying the first test for a zero difference in means to

$$(x_{11} - \Delta_0, x_{12} - \Delta_0, \ldots, x_{1n_1 - \Delta_0}), (x_{21}, x_{22}, \ldots, x_{2n_2}).$$

## Description of the Procedure

1. Rank all the observations from smallest to largest into one list with $N = n_1 + n_2$ observations.
2. Let $r_j$ be the rank of the observation $x_{1j}$.
3. Set
$$W = \sum_{j=1}^{n_1} r_j$$
which is the sum of the ranks associated with the data from the $x_1$-population.

If $H_0$ is true, then each overall ranking of the $N$ combined observations would have the same probability, but if, for instance,

$$H_a : \mu_1 - \mu_2 > 0$$

then the ranks contributing to $W$, arising from the $x_1$-population with the larger mean, would be larger than expected under $H_0$, leading to an observed $W$ value substantially above the mean. The Wilcoxon rank sum test therefore rejects $H_0$ if $W$ is beyond some reasonable value. This test is sometimes called the Mann-Whitney test because it was originally derived as an alternate, but equivalent, formulation by H. B. Mann and D. R. Whitney in 1947.

The exact distribution of $W$ has been tabulated, but we will base our procedure on the normal approximation. It can be shown that if $H_0$ is true, then

$$\mu_W = \text{mean of } W = \frac{n_1(n_1 + n_2 + 1)}{2}$$

$$\sigma_w = \text{standard deviation of } W = \sqrt{\frac{n_1 n_2 (n_1 + n_2 + 1)}{12}}$$

Calculate

$$z = \frac{W - \mu_W}{\sigma_W}.$$

For a fixed level $\alpha$ test:

$$\begin{array}{ll} \text{if } H_a : \mu_1 - \mu_2 > 0 & \text{reject } H_0 \text{ if } z > z^*_\alpha \\ \text{if } H_a : \mu_1 - \mu_2 < 0 & \text{reject } H_0 \text{ if } z < -z^*_\alpha \\ \text{if } H_a : \mu_1 - \mu_2 \neq 0 & \text{reject } H_0 \text{ if } |z| > z^*_{\frac{\alpha}{2}} \end{array}$$

where $z^*$ represents the corresponding upper critical value of a standard normal distribution.

## The Wilcoxon Rank Sum Test in Practice

We describe the steps required for carrying out the Wilcoxon rank sum test.

**Example 12.1.** (Example 12.1 in text.) Does the presence of small numbers of weeds reduce the yield of corn? Lamb's-quarter is a common weed in corn fields. A researcher planted corn at the same rate in eight small plots of ground, then weeded the corn rows by hand to allow no weeds in four randomly selected plots and exactly three lamb's-quarter plants per meter of row in the other four plots. Here are the yields of corn (bushels per acre) in each of the plots.

| Weeds per meter | Yield (bu/acre) | | | |
|---|---|---|---|---|
| 0 | 166.7 | 172.2 | 165.0 | 176.9 |
| 3 | 158.6 | 176.4 | 153.1 | 156.0 |

Normal quantile plots suggest that the data may be right-skewed. The samples are too small to assess normality adequately or to rely on the robustness of the two-sample $t$ test. We may prefer to use a test that does not require normality. Carry out the significance test

$$H_0 : \text{ no difference in distribution of yields}$$
$$H_a : \text{ yields are systematically higher in weed-free plots}$$

**Solution.** Figure 12.1 shows the Excel formulas required and the corresponding values taken when applied to the above data set.

1. Enter the labels "Sample 1," "Sample 2," "Population," "Combined," and "Rank" in cells A3:E3.

2. Record the observations in Sample 1 in cells A4:A7 and copy them to D4:D7. Record the observations in Sample 2 in cells B4:B7 and copy them to D8:D11. Enter the value 1 in C4:C7, the value 2 in C8:C11, and, finally, enter the values $\{1, 2, 3, 4, 5, 6, 7, 8\}$ in E4:E11.

3. **Name** the ranges for Sample 1, Sample 2, Population, Combined, and Rank.

4. Rank the combined data set that has been copied into D4:D11, and carry their corresponding sample numbers (Sample 1 = 0 weeds per meter, Sample 2 = 3 weeds per meter) back to C4:C11 as follows. Select C3:D11 and from the Menu Bar choose **Data** – **Sort** to bring up the **Sort** dialog box shown in Figure 12.2. Click OK. This ranks the combined sample in D4:D11 in increasing order and also carries the corresponding sample labels in C4:C11. Figure 12.1 shows the results of the sort.

| | A | B | C | D | E | F | G | H |
|---|---|---|---|---|---|---|---|---|
| 1 | | | | *Wilcoxon Rank Sum Test* | | | | |
| 2 | | | | | | | | |
| 3 | Sample1 | Sample2 | Population | Combined | Rank | | | |
| 4 | 166.7 | 158.6 | 2 | 153.1 | 1 | | | |
| 5 | 172.2 | 176.4 | 2 | 156.0 | 2 | | | |
| 6 | 165.0 | 153.1 | 2 | 158.6 | 3 | | | |
| 7 | 176.9 | 156.0 | 1 | 165.0 | 4 | | | |
| 8 | | | 1 | 166.7 | 5 | | | |
| 9 | | | 1 | 172.2 | 6 | | | |
| 10 | | | 2 | 176.4 | 7 | | | |
| 11 | | | 1 | 176.9 | 8 | | | |
| 12 | | | | | | | | |
| 13 | User Input | | | | | | | |
| 14 | alpha | 0.05 | | | | | | |
| 15 | Summary Statistics | | | | | | | |
| 16 | n_1 | 4 | | =COUNT(Sample1) | | | | |
| 17 | n_2 | 4 | | =COUNT(Sample2) | | | | |
| 18 | Calculations | | | | | | | |
| 19 | W | 23 | | =SUMIF(Population, "=1", Rank) | | | | |
| 20 | mu | 18 | | =n_1*(n_1+n_2+1)/2 | | | | |
| 21 | sigma | 3.464 | | =SQRT(n_1*n_2*(n_1+n_2+1)/12) | | | | |
| 22 | z | 1.443 | | =(W-mu)/sigma | | | | |
| 23 | Lower Test | | | | | | | |
| 24 | lower_z | -1.645 | | =NORMSINV(alpha) | | | | |
| 25 | Decision | Do Not Reject H0 | | =IF(z<lower_z,"Reject H0","Do Not Reject H0") | | | | |
| 26 | Pvalue | 0.926 | | =NORMSDIST(z) | | | | |
| 27 | Upper Test | | | | | | | |
| 28 | upper_z | 1.645 | | =-NORMSINV(alpha) | | | | |
| 29 | Decision | Do Not Reject H0 | | =IF(z>upper_z,"Reject H0","Do Not Reject H0") | | | | |
| 30 | Pvalue | 0.0745 | | =1-NORMSDIST(z) | | | | |
| 31 | Two-Sided Test | | | | | | | |
| 32 | two_z | 1.960 | | =ABS(NORMSINV(alpha/2)) | | | | |
| 33 | Decision | Do Not Reject H0 | | =IF(ABS(z)>two_z,"Reject H0","Do Not Reject H0") | | | | |
| 34 | Pvalue | 0.149 | | =2*(1-NORMSDIST(ABS(z))) | | | | |

Figure 12.1: Wilcoxon Rank Sum Test – Formulas and Values

5. Next we carry out calculations akin to those in Chapter 6 (see Figure 6.3) for a one-sample test of a normal mean. Enter the labels as shown in cells A13:A34 on Figure 12.1, and **Name** the corresponding ranges in the respective column B cells to be able to refer to $n\_1$, $n\_2$, $W$, mu, sigma, $z$, lower_$z$, upper_$z$, and two_$z$ by name in the ensuing formulas shown. The formulas to be entered in column B are presented in column D, and the values taken when the formulas are applied to this data (that is, what you will see in your workbook) are shown in the corresponding rows in column B. For instance, enter

$$= \texttt{SUMIF}(\text{Population}, = 1, \text{Rank})$$

in cell B19. This Excel function adds those cells under Rank whose corresponding Population is 1. In other words, the function adds the ranks corresponding to observations from Sample 1. The answer 23 appears in cell B19 and is the value of the Wilcoxon rank sum statistic. Formulas for all three types of alternate hypotheses are provided in Figure 12.1. Use only the one appropriate for the problem at hand.

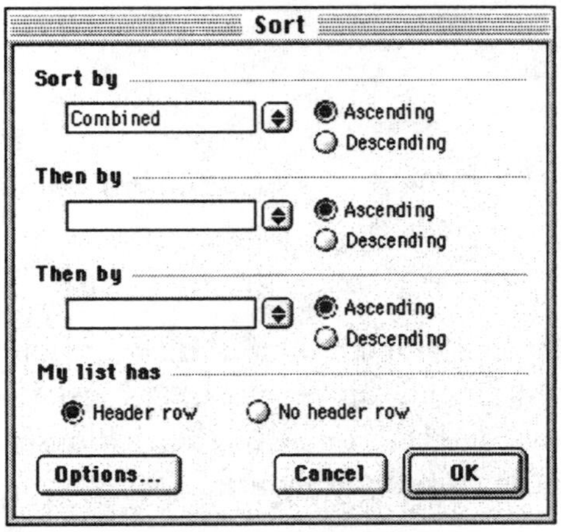

Figure 12.2: Sort Dialog Box

## Interpreting the Results

We read off

$$\begin{aligned} W &= 23 && \text{(cell B19)} \\ \mu_W &= 18 && \text{(cell B20)} \\ \sigma_W &= 3.464 && \text{(cell B21)} \\ z-\text{statistic} &= 1.443 && \text{(cell B22)} \end{aligned}$$

The alternate hypothesis is

$$H_a : \mu_1 - \mu_2 > 0$$

so that rows 27–30 are appropriate (rows 23–26 for a lower-tailed test and rows 31–34 for a two-tailed test). We find that

$$\begin{aligned} \text{upper critical value } z^* &= 1.645 && \text{(cell B28)} \\ \text{decision rule} &= \text{“Do not reject”} && \text{(cell B29)} \\ P-\text{value} &= 0.0745 && \text{(cell B30)} \end{aligned}$$

We conclude that the data is not significant at the nominal 5% level of significance.

## Continuity Correction for the Normal Approximation

A more accurate $P$-value is obtained by applying the continuity correction, which adjusts for the fact that a continuous distribution (the normal) is being used to approximate a discrete distribution W. If we use a correction of 0.5, then the upper-tailed test statistic becomes

$$z = \frac{W - 0.5 - \mu_W}{\sigma_W}$$

and this leads to

$$z = 1.299$$

and

$$P - \text{value} = 1 - \Phi(1.299) = 0.0970.$$

### Ties

Theoretically, the assumption of a continuous distribution ensures that all $n_1 + n_2$ observed values will be different. In practice, ties are sometimes observed. The common practice is to average the ranks for the tied observations and carry on as above with a change in the standard deviation. Use

$$\sigma_W^2 = \frac{n_1 n_2}{12}\left(n_1 + n_2 + 1 - \frac{\sum_{i=1}^{G} t_i(t_i^2 - 1)}{(n_1 + n_2)(n_1 + n_2 - 1)}\right)$$

where $G$ is the number of tied groups and $t_i$ are the number of tied observations in the $i$th tied group. Unless $G$ is large, the adjustment in the formula for the variance makes little difference.

## 12.2 The Wilcoxon Signed Rank Test

The Wilcoxon signed rank test is a nonparametric version of the one-sample procedures based on the assumption of normal population discussed in Chapters 6 and 7. The key assumption is that the data arise from a population symmetric about its mean.

For *this* reason one of its most useful applications is in a matched-pairs setting with $n$ pairs $(x_{1i}, x_{2i})$ of observations, where it is natural to assume that the populations from which the pairs are taken differ only by a shift in the mean (that is, the population distribution shapes are otherwise the same). The differences then satisfy the requirement of symmetry under the null hypothesis of equality of means.

### Description of the Matched-Pairs Procedure

The data consist of $n$ pairs $(x_{1i}, x_{2i})$ of observations. The $\{x_{1i}\}$ are a sample from a population with mean $\mu_1$ and the $\{x_{2i}\}$ are a sample from a population with mean $\mu_2$. The null hypothesis is

$$H_0 : \mu_1 - \mu_2 = 0$$

1. Form the absolute differences $|d_j|$, where $d_j = x_{1j} - x_{2j}$.

2. Let $r_j$ be the rank of $|d_j|$ in the joint ranking of the $\{|d_j|\}$, from smallest to largest.

3. Form the sum of the positive signed ranks

$$S_+ = \sum r_j$$

where the sum is taken over all ranks $r_j$ for which the corresponding difference $d_j$ is positive.

The Wilcoxon signed rank procedure rejects $H_0$ if $S_+$ is beyond some reasonable value, in particular for values of $S_+$ that are too large or too small.

As with the Wilcoxon rank sum statistic $W$, there exist tables of the exact distribution of $S_+$, but we will base our procedure on the normal approximation. When $H_0$ is true the mean and the standard deviation of $S_+$ are given by

$$\mu_{S_+} = \frac{n(n+1)}{4}$$

$$\sigma_{S_+} = \sqrt{\frac{n(n+1)(2n+1)}{24}}$$

We then calculate

$$z = \frac{S_+ - \mu_{S_+}}{\sigma_{S_+}}$$

and for a fixed level $\alpha$ test:

$$\begin{array}{ll} \text{if } H_a : \mu_1 - \mu_2 > 0 & \text{reject } H_0 \text{ if } z > z^*_\alpha \\ \text{if } H_a : \mu_1 - \mu_2 < 0 & \text{reject } H_0 \text{ if } z < -z^*_\alpha \\ \text{if } H_a : \mu_1 - \mu_2 \neq 0 & \text{reject } H_0 \text{ if } |z| > z^*_{\frac{\alpha}{2}} \end{array}$$

## The Wilcoxon Signed Rank Test in Practice

We illustrate the Wilcoxon signed rank test for matched pairs.

**Example 12.2.** (Example 12.8 in text.) A study of early childhood education asked kindergarten students to tell two fairy tales that had been read to them earlier in the week. The first tale had been read to them and the second had been read but also illustrated with pictures. An expert listened to a recording of the children and assigned a score for certain uses of language. Here are the data for five "low progress" readers in a pilot study:

| Child | 1 | 2 | 3 | 4 | 5 |
|---|---|---|---|---|---|
| Story 2 | 0.77 | 0.49 | 0.66 | 0.28 | 0.38 |
| Story 1 | 0.40 | 0.72 | 0.00 | 0.36 | 0.55 |
| Difference | 0.37 | −0.23 | 0.66 | −0.08 | −0.17 |

We wonder if illustrations improve how the children retell a story. We would like to test the hypotheses

$$H_0 : \text{ scores have the same distribution for both stories}$$
$$H_a : \text{ scores are systematically higher for story 2.}$$

Because this is a matched-pairs design, we base our inference on the differences. The matched-pairs $t$ test gives $t = 0.635$ with a one-sided $P$-value of 0.280. As displays of the data suggest a mild lack of normality, carry out a nonparametric significance test.

**Solution.** Figure 12.3 shows the Excel formulas required and the corresponding values taken for the above data set. The calculations are similar to those in the previous section.

| | A | B | C | D | E | F | G |
|---|---|---|---|---|---|---|---|
| 1 | *Wilcoxon Signed Rank Test - Matched Pairs* | | | | | | |
| 2 | | | | | | | |
| 3 | Sample1 | Sample2 | Diff | | Ranked_Diff | Ranked_Abs_Diff | Rank |
| 4 | 0.77 | 0.40 | 0.37 | | -0.08 | 0.08 | 1 |
| 5 | 0.49 | 0.72 | -0.23 | | -0.17 | 0.17 | 2 |
| 6 | 0.66 | 0.00 | 0.66 | | -0.23 | 0.23 | 3 |
| 7 | 0.28 | 0.36 | -0.08 | | 0.37 | 0.37 | 4 |
| 8 | 0.38 | 0.55 | -0.17 | | 0.66 | 0.66 | 5 |
| 9 | | | | | | | |
| 10 | | | | | | | |
| 11 | User Input | | | | | | |
| 12 | alpha | 0.05 | | | | | |
| 13 | Summary Statistics | | | | | | |
| 14 | n | 5 | | =COUNT(Sample1) | | | |
| 15 | Calculations | | | | | | |
| 16 | Splus | 9 | | =SUMIF(Ranked_Diff, ">0", Rank) | | | |
| 17 | mu | 7.5 | | =n*(n+1)/4 | | | |
| 18 | sigma | 3.708 | | =SQRT(n*(n+1)*(2*n+1)/24) | | | |
| 19 | z | 0.405 | | =(Splus-mu)/sigma | | | |
| 20 | Lower Test | | | | | | |
| 21 | lower_z | -1.645 | | =NORMSINV(alpha) | | | |
| 22 | Decision | Do Not Reject H0 | | =IF(z<lower_z,"Reject H0","Do Not Reject H0") | | | |
| 23 | Pvalue | 0.657 | | =NORMSDIST(z) | | | |
| 24 | Upper Test | | | | | | |
| 25 | upper_z | 1.645 | | =-NORMSINV(alpha) | | | |
| 26 | Decision | Do Not Reject H0 | | =IF(z>upper_z,"Reject H0","Do Not Reject H0") | | | |
| 27 | Pvalue | 0.3429 | | =1-NORMSDIST(z) | | | |
| 28 | Two-Sided Test | | | | | | |
| 29 | two_z | 1.960 | | =ABS(NORMSINV(alpha/2)) | | | |
| 30 | Decision | Do Not Reject H0 | | =IF(ABS(z)>two_z,"Reject H0","Do Not Reject H0") | | | |
| 31 | Pvalue | 0.686 | | =2*(1-NORMSDIST(ABS(z))) | | | |

Figure 12.3: Wilcoxon Signed Rank – Matched Pairs

1. Enter the labels "Sample 1," "Sample 2," and "Diff" in cells A3:C3, and the labels "Ranked_Diff," "Ranked_Abs_Diff," and "Rank" in cells E3:G3.

2. Record the observations in Sample 1 in cells A4:A8 and the observations for Sample 2 in cells B4:B8. In cells C4:C8 record the difference between Sample 1 and Sample 2. Enter the values $\{1, 2, 3, 4, 5\}$ in G4:G8.

3. Next **Name** the ranges for the corresponding labels Sample 1, Sample 2, Ranked_Diff, Ranked_Abs_Diff, and Rank to include the respective cells in rows 4–8.

4. In a different part of the sheet, copy the values of the differences (not their formulas, if Excel calculated them) and then enter their absolute values using the Excel function `ABS()`, which gives the absolute value of its argument. As in the corresponding step 4 of the previous section, rank the absolute values of the differences in increasing order and carry along the actual differences. Copy the results to cells E4:F8 as shown in Figure 12.3, showing the ranked absolute differences in column F and the actual differences in column E . We need the latter to recognize which absolute differences correspond to positive differences in calculating $S_+$.

5. The calculations required are shown in the lower portion of Figure 12.3, where we have included in column D the formulas that are to be entered in column B. The actual values taken by these formulas, and which will appear on your workbook, are in column B. Refer to step 5 of the previous section for the analogous details, and remember to name all ranges used in the formulas shown.

## Interpreting the Results

We read off

$$\begin{aligned} S_+ &= 9 && \text{(cell B16)} \\ \mu_{S_+} &= 7.5 && \text{(cell B17)} \\ \sigma_{S_+} &= 3.708 && \text{(cell B18)} \\ z-\text{statistic} &= 0.405 && \text{(cell B19)} \end{aligned}$$

The alternate hypothesis requires an upper-tailed test for which

$$P-\text{value} = 0.3429 \qquad \text{(cell B27)}$$

We conclude that the data is not significant.

## Continuity Correction for the Normal Approximation

As with the Wilcoxon rank sum test, a more accurate $P$-value is obtained with the continuity correction

$$z = \frac{S_+ - 0.5 - \mu_{S_+}}{\sigma_{S_+}}$$

and this leads to

$$z = 0.270$$

$$P-\text{value} = 1 - \Phi(0.270) = 0.3937.$$

### Ties and Zero Values

If there are zeros among the differences $\{d_i\}$, discard them and use for $n$ the number of non-zero $\{d_i\}$. If there are any ties, then use the average rank for each set of tied observations and apply the procedure with variance

$$\sigma_W^2 = \frac{1}{24}\left(n(n+1)(2n+1) - \frac{\sum_{i=1}^{G} t_i(t_i^2-1)}{2}\right)$$

where $G$ is the number of tied groups and $t_i$ are the number of tied observations in the $i$th tied group.

## 12.3 The Kruskal-Wallis Test

In this section we generalize the Wilcoxon rank sum test to situations involving independent samples from I populations when the assumptions required for validity of the one-way ANOVA in Chapter 10 cannot be substantiated.

The data consist of $N = \sum_{i=1}^{I} n_i$ observations with $n_i \geq 1$ observations $\{x_{ij} : 1 \leq j \leq n_i\}$ taken from population $i$. The assumption replacing normality is

$$x_{ij} = \mu_i + \varepsilon_{ij} \quad 1 \leq i \leq I,\ 1 \leq j \leq n_i$$

where the errors $\{\varepsilon_{ij}\}$ are mutually independent with mean 0 and have the same *continuous* distribution. If we let $F(x)$ be the cumulative distribution function (c.d.f.) of a generic error term, this is tantamount to

$$F_i(x) \equiv F(x - \mu_i) \quad 1 \leq i \leq I$$

being the c.d.f. of population $i$.

The significance test is

$$H_0 : \mu_1 = \mu_2 = \cdots = \mu_I$$
$$H_a : \text{ not all of the } \mu_i \text{ are equal,}$$

and the procedure generalizing the rank sum test is called the Kruskal-Wallis test.

### Description of the Procedure

1. Rank all the observations jointly from smallest to largest.

2. Let $r_{ij}$ be the rank of observation $x_{ij}$.

3. Set
$$R_i = \sum_{j=1}^{n_i} r_{ij}$$
which is the sum of the ranks associated with sample $i$.

Denote by

$$\bar{R}_i = \frac{1}{n_i} R_i$$

the average rank in sample $i$. If $H_0$ is true, then by symmetry the mean of any rank $r_{ij}$ is $E(r_{ij}) = \frac{N+1}{2}$, which is the average of the integers $\{1, 2, \ldots, N\}$ and therefore $E[\bar{R}_i] = E\left[\frac{1}{n_i} R_i\right] = E\left[\frac{1}{n_i} \sum_{j=1}^{n_i} r_{ij}\right] = \frac{N+1}{2}$. Thus, we would expect the ranks to be uniformly intermingled among the I samples. But if $H_0$ is false, then some samples will tend to have many small ranks, while others will have many large ranks. Just as in ANOVA, we take the sum of squares of the differences between the average rank $\bar{R}_i$ of each sample and the overall average $\frac{N+1}{2}$ by computing

$$\frac{12}{N(N+1)} \sum_{i=1}^{I} n_i \left(\bar{R}_i - \frac{N+1}{2}\right)^2$$

which can be expressed equivalently as

$$H = \frac{12}{N(N+1)} \sum_{i=1}^{I} \frac{R_i^2}{n_i} - 3(N+1)$$

and called the Kruskal-Wallis statistic. We then reject $H_0$ for "large" values of $H$.

Tables of critical values exist for small values of the $\{n_i\}$, but it is customary to use a normal approximation, which provides an approximate sampling distribution

$H$ is approximately chi-square with $I - 1$ degrees of freedom.

Therefore the test is:

$$\text{Reject } H_0 \text{ if } H > \chi^2_\alpha$$

where $\chi^2_\alpha$ is the upper critical $\alpha$ value of a chi-square distribution on $I - 1$ degrees of freedom.

## The Kruskal-Wallis Test in Practice

**Example 12.3.** (Example 12.13 in text.) Lamb's-quarter is a common weed that interferes with the growth of corn. A researcher planted corn at the same rate in 16 small plots of ground, then randomly assigned the plots to four groups. He weeded the plots by hand to allow a fixed number of lamb's-quarter plants to grow in each meter of corn row. These numbers were 0, 1, 3, and 9 in the four groups of plots. No other weeds were allowed to grow, and all plots received identical treatment, except for the weeds. Here are the yields of corn (bushels per acre) in each of the plots.

| Weeds per meter | Corn yield | Weeds per meter | Corn yield | Weeds per meter | Corn yield | Weeds per meter | Corn yield |
|---|---|---|---|---|---|---|---|
| 0 | 166.7 | 1 | 166.2 | 3 | 158.6 | 9 | 162.8 |
| 0 | 172.2 | 1 | 157.3 | 3 | 176.4 | 9 | 142.4 |
| 0 | 165.0 | 1 | 166.7 | 3 | 153.1 | 9 | 162.7 |
| 0 | 176.9 | 1 | 161.1 | 3 | 156.0 | 9 | 162.4 |

The summary statistics are:

| Weeds | $n$ | Mean | Std Dev |
|---|---|---|---|
| 0 | 4 | 170.200 | 5.422 |
| 1 | 4 | 162.825 | 4.469 |
| 3 | 4 | 161.025 | 10.498 |
| 9 | 4 | 157.575 | 10.118 |

The sample standard deviations do not satisfy our rule of thumb that for safe use of ANOVA the largest should not exceed twice the smallest. Normal quantile plots show that outliers are present in the yields for 3 and 9 weeds per meter. Use the Kruskal-Wallis procedure to test:

$H_0$ : yields have the same distribution in all groups

$H_a$ : yields are systematically higher in some groups than others.

**Solution.** Figure 12.4 shows the Excel formulas required and the corresponding values taken when applied to the above data set.

As we have already described in detail the construction of the two earlier nonparametric procedures, we leave as an exercise to the reader the duplication of the above workbook. Beginning in row 9, column C shows the formulas to be entered into the adjacent cells of column B, where the numerical evaluation of the formulas is shown.

## Interpreting the Results

We read off

$$\begin{aligned} H &= 5.36 \quad \text{(cell B25)} \\ \chi^2_{.05} &= 7.815 \quad \text{(cell B26)} \\ P-\text{value} &= 0.147 \quad \text{(cell B27)} \end{aligned}$$

The data is not significant, and there is no convincing evidence that more weeds decrease yield.

| | A | B | C | D | E | F | G | H |
|---|---|---|---|---|---|---|---|---|
| 1 | | | | *Kruskal-Wallis Test* | | | | |
| 2 | | | | | | | | |
| 3 | Sample1 | Sample2 | Sample3 | Sample4 | | Pop | Combined | Rank |
| 4 | 166.7 | 166.2 | 158.6 | 162.8 | | 4 | 142.4 | 1 |
| 5 | 172.2 | 157.3 | 176.4 | 142.4 | | 3 | 153.1 | 2 |
| 6 | 165.0 | 166.7 | 153.1 | 162.7 | | 3 | 156.0 | 3 |
| 7 | 176.9 | 161.1 | 156.0 | 162.4 | | 2 | 157.3 | 4 |
| 8 | | | | | | 3 | 158.6 | 5 |
| 9 | R_1 | 52 | =SUMIF(Pop, "=1", Rank) | | | 2 | 161.1 | 6 |
| 10 | R_2 | 34 | =SUMIF(Pop, "=2", Rank) | | | 4 | 162.4 | 7 |
| 11 | R_3 | 25 | =SUMIF(Pop, "=3", Rank) | | | 4 | 162.7 | 8 |
| 12 | R_4 | 25 | =SUMIF(Pop, "=4", Rank) | | | 4 | 162.8 | 9 |
| 13 | | | | | | 1 | 165.0 | 10 |
| 14 | n_1 | 4 | =COUNT(A4:A7) | | | 2 | 166.2 | 11 |
| 15 | n_2 | 4 | =COUNT(B4:B7) | | | 1 | 166.7 | 12 |
| 16 | n_3 | 4 | =COUNT(C4:C7) | | | 2 | 166.7 | 13 |
| 17 | n_4 | 4 | =COUNT(D4:D7) | | | 1 | 172.2 | 14 |
| 18 | N | 16 | =n_1+n_2+n_3+n_4 | | | 3 | 176.4 | 15 |
| 19 | | | | | | 1 | 176.9 | 16 |
| 20 | | 676 | =R_1^2/n_1 | | | | | |
| 21 | | 289 | =R_2^2/n_2 | | | | | |
| 22 | | 156.25 | =R_3^2/n_3 | | | | | |
| 23 | | 156.25 | =R_4^2/n_4 | | | | | |
| 24 | | | | | | | | |
| 25 | H | 5.360294 | =(12/(N*(N+1)))*SUM(B21:B24) - 3*(N+1) | | | | | |
| 26 | Critical 5% | 7.815 | =CHIINV(0.05,3) | | | | | |
| 27 | P-value: | 1.47E-01 | =CHIDIST(H,3) | | | | | |

Figure 12.4: Kruskal-Wallis Test – Formulas and Values

## Ties

We have ignored a tie in the above calculation; the value 166.7 appears in Sample 1 and in Sample 2. For a more accurate calculation, we give to the value 166.7 the average rank 12.5. Then replace $H$ with

$$H' = \frac{H}{1 - \sum_{i=1}^{G} \frac{t_i(t_i^3-1)}{N^3-N}}$$

where $G$ is the number of tied groups and $t_i$ are the number of tied observations in the $i$th tied group.

> **Exercise.** Replace the ranks of the two observations 166.7 in your workbook with common rank 12.5 and then calculate $H'$ as
>
> $$H' = \frac{H}{1 - \frac{2(2^3-1)}{16^3-16}} = \frac{H}{.9965686}$$
>
> Show that the $P$-value becomes 0.134.

# Appendix – Excel 5/95

With respect to data analysis capabilities, the releases of Excel 97 (Windows) and Excel 98 (Macintosh) corrected some bugs in the Data Analysis Toolpak and changed the interface to the construction of charts and formulas. Components of some dialog boxes were juxtaposed or shifted (horizontally or vertically). Otherwise few substantive changes were made that affect use of Excel in statistics.

Excel 97/98 has been used as the basis this book. Still there is a large base of Excel 5/95 users, both home and university. Also, a student may be exposed to one version at school, another at home, and possibly a third at work.

An instructor using this manual could make the adjustments for Excel 5/95 in using this manual. Nonetheless in order to provide a ramp into Excel 5/95, this appendix has been provided covering all changes that are relevant, mainly material in Chapters 1 and 2. Remarkably, it is only in these opening chapters on data description and functions where a detailed separate exposition is required. Elsewhere in the book, where Excel 5/95 and Excel 97/98 differ, either because of dialog boxes or different pull-down menus, for instance, parallel step-by-step descriptions are indicated for each version. There are only a few techniques which require this, and in each case the difference requires but a few lines.

Figures in this Appendix were taken using Excel 5 on a Macintosh.

## A.1 The ChartWizard

The **ChartWizard** is a step-by-step approach to creating informative graphs. In Excel 5/95 a sequence of five dialog boxes guides the user through the creation of a customized chart. The user provides details about the chart type, formatting, titles, legends, etc. The **ChartWizard** can be activated either from the button on the **Standard Toolbar** or by choosing **Insert – Chart** from the Menu Bar. If you embed the chart on the workbook the mouse pointer becomes a cross hair + with the image of a chart next to it. Click on the workbook to locate the upper left corner of your graph output to create a default size. Otherwise click and drag the cursor to the lower right corner to create the desired size chart.

We illustrate use of the ChartWizard with the same example (Example 1.1) from Chapter 1.

| | A | B |
|---|---|---|
| 1 | *Cause* | *Count* |
| 2 | auto | 43363 |
| 3 | falls | 10483 |
| 4 | drowning | 4350 |
| 5 | fires | 4235 |
| 6 | poison | 9072 |
| 7 | other | 18899 |

Figure A.1: Causes of Accidental Deaths

**Example A.1.** (Exercise 1.4 page 8 in text.) Display the data in the workbook in Figure A.1 showing the causes of accidental deaths in the United States in 1995 as a bar graph.

## Creating a Bar Chart

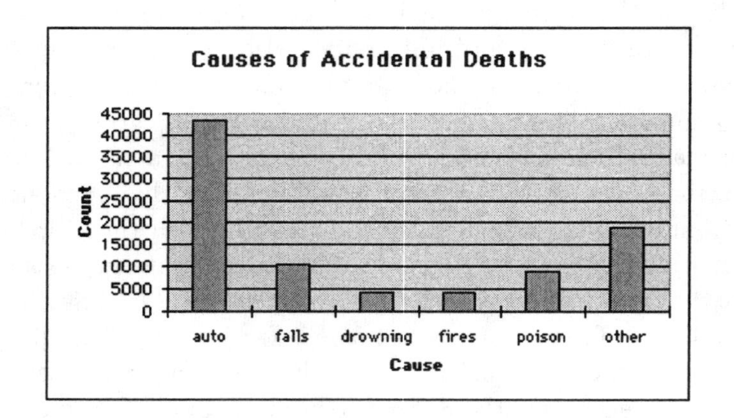


Figure A.2: Excel Bar Chart

The steps for creating a bar graph are given below. For other types of graphical displays, make appropriate choices from the same sequence of dialog boxes.

Figure A.2 is a bar graph produced by Excel which displays the same information as in Figure A.1. The following steps describe how it is obtained. First enter the data and labels in cells A1:B7 and format the display as in Figure A.1 for presentation purposes.

1. Step 1. Select the cells where you have located the data, in this case cells A2:B7, and click on the **ChartWizard** on the **Standard Toolbar**. The pointer changes to a cross hair +. Now click in cell A10 (or some other cell) to locate the chart. (Alternatively you may first click on the **ChartWizard**, then click in cell A10 to locate the chart, and finally click and drag over the range A2:B7.) In either case, dialog box 1 of 5 appears (Figure A.3) with the

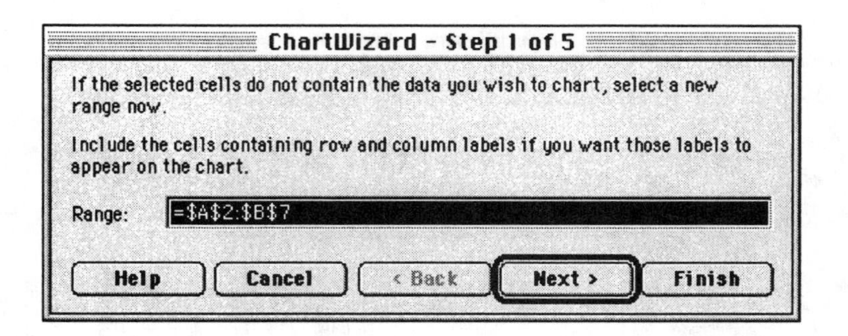


Figure A.3: Chart Wizard Step 1

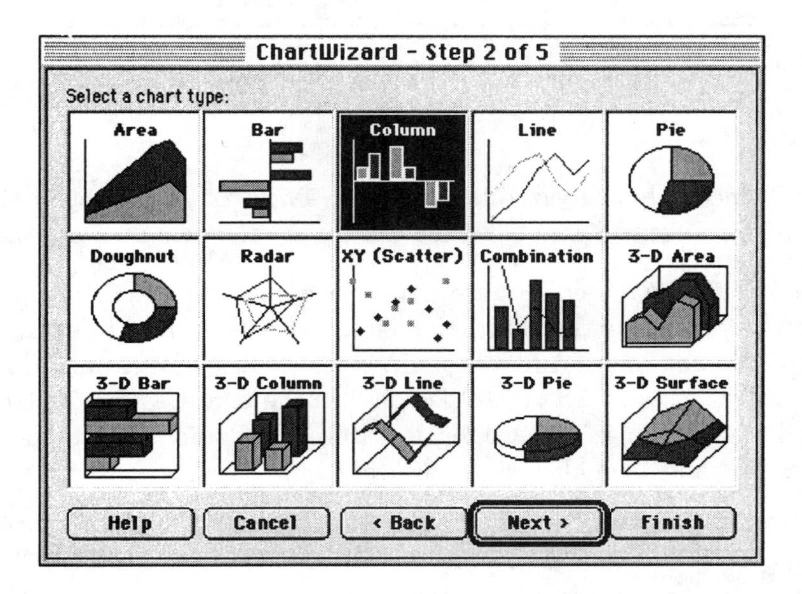


Figure A.4: Chart Wizard Step 2

selected range highlighted. Modify the range if necessary and then confirm by clicking the Next button.

2. Step 2. The **ChartWizard** (Figure A.4) displays the types of graphs that are available. Select **Column** chart and click Next.

3. Step 3. Various formatting options (Figure A.5) available for the chart type selected in Step 2 are presented. Select Format **6** and click Next.

4. Step 4. This dialog box (Figure A.6) presents the chart as it will appear by default. You may confirm or change its appearance. Because each variable is located in a column, select the radio button **Columns**. The first column of the data set will be located on the X axis. Therefore "1" should be in the text area for the Column(s) for Category(X) axis labels. We don't require a

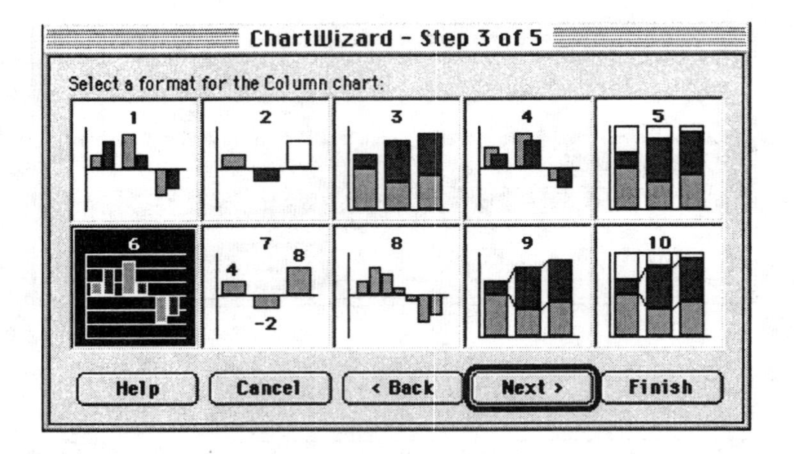


Figure A.5: Chart Wizard Step 3

legend since only one variable is plotted. Thus "0" should be in the box for Row(s) for Legend Text. All these choices are the defaults here. Click Next.

5. Step 5. The final step (Figure A.7) allows the user to customize the chart by adding titles for the chart, the axes, and a legend. By default the radio button for **Yes** under Add a legend? is selected. **Change** this to **No**. Enter the text "Causes of Accidental Deaths" in the text area for the Chart Title. Finally, type "Cause" for Axis Title Category (X) and "Count" for Axis Title Value (Y), and then click the Finish button.

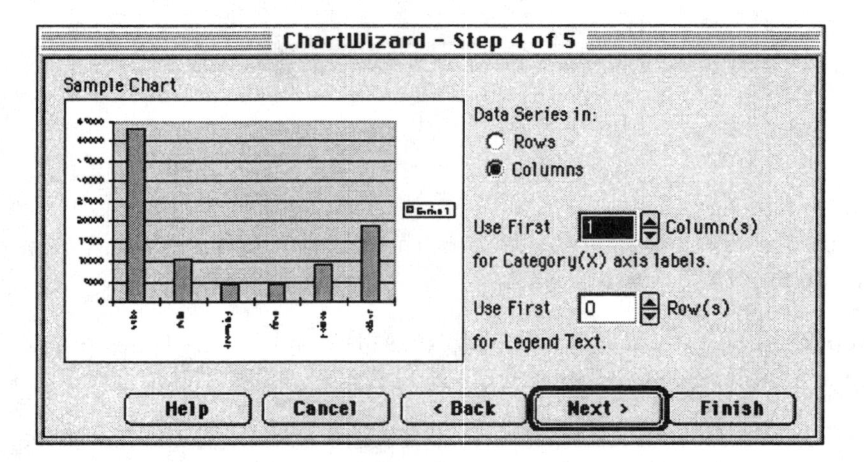


Figure A.6: Chart Wizard Step 4

6. The chart appears with **eight handles** (Figure A.8) indicating that it is selected. The chart can be resized by selecting a handle and then dragging the handle to the desired size. The chart can also be moved. Click the

interior of the chart and drag it to another location (holding the mouse button down). Then click outside the chart to deselect.

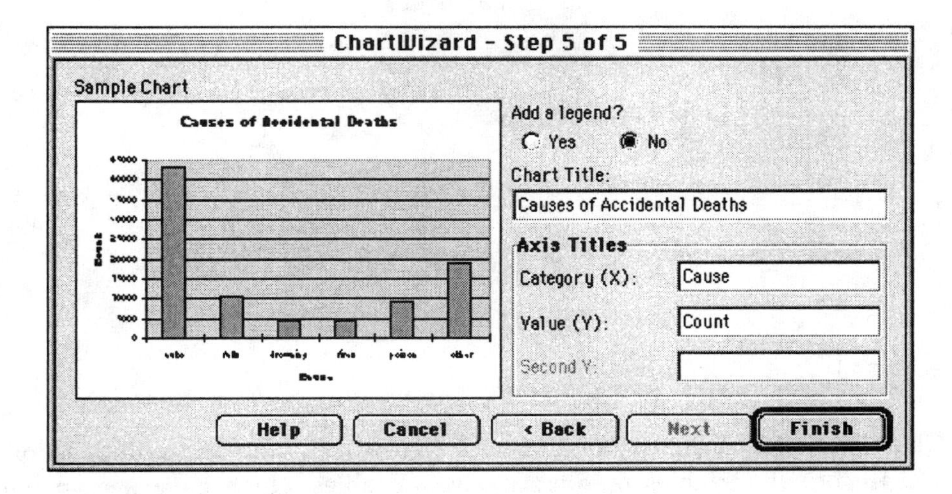


Figure A.7: Chart Wizard Step 5

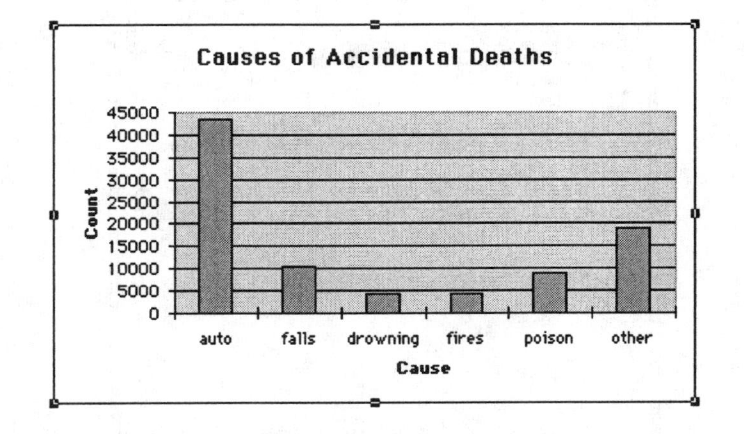


Figure A.8: Bar Chart Selected

## Pie Charts

To produce a pie chart as in Figure A.9 select **Pie** in place of Column in **Step 2** of the ChartWizard and follow the remaining steps.

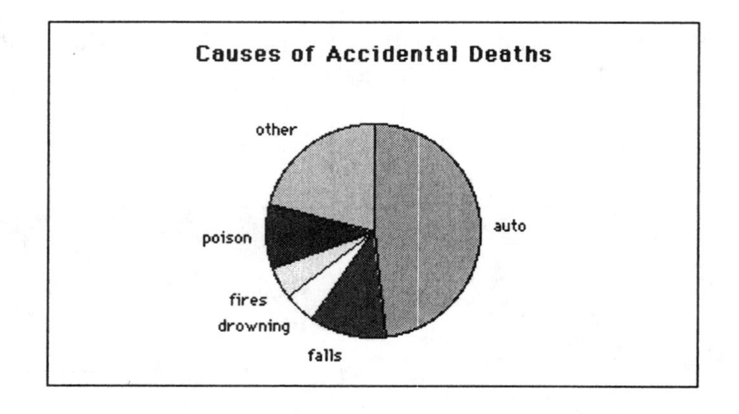


Figure A.9: Pie Chart

## A.2 Histograms

The ChartWizard is designed for use with data that is already grouped, for instance, categorical variables or quantitative variables which have been grouped into categories or intervals. For raw numerical data, Excel provides additional commands using the **Analysis ToolPak**.

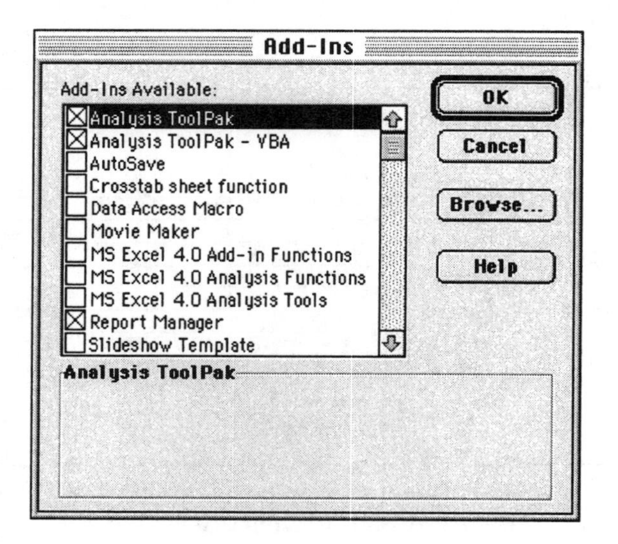


Figure A.10: Add-Ins Dialog Box

To determine whether this toolpak is installed, choose **Tools – Add-Ins** from the Menu Bar for the **Add-Ins** dialog box (Figure A.10). Depending on whether other Add-Ins have been loaded, your box might appear slightly different. If the Analysis ToolPak box is not checked, then select it and click OK. It will now be an option in the pull-down menu when you choose **Tools – Data Analysis**.

| | A | B | C | D | E | F | G | H | I | J | K |
|---|---|---|---|---|---|---|---|---|---|---|---|
| 1 | *Survival Times of Guinea Pigs* | | | | | | | | | | |
| 2 | 43 | 45 | 53 | 56 | 56 | 57 | 58 | 66 | 67 | 73 | |
| 3 | 74 | 79 | 80 | 80 | 81 | 81 | 81 | 82 | 83 | 83 | |
| 4 | 84 | 88 | 89 | 91 | 91 | 92 | 92 | 97 | 99 | 99 | |
| 5 | 100 | 100 | 101 | 102 | 102 | 102 | 103 | 104 | 107 | 108 | |
| 6 | 109 | 113 | 114 | 118 | 121 | 123 | 126 | 128 | 137 | 138 | |
| 7 | 139 | 144 | 145 | 147 | 156 | 162 | 174 | 178 | 179 | 184 | |
| 8 | 191 | 198 | 211 | 214 | 243 | 249 | 329 | 380 | 403 | 511 | |
| 9 | 522 | 688 | | | | | | | | | |

Figure A.11: Survival Times of Guinea Pigs

## Histogram from Raw Data

**Example A.2.** (Exercise 1.80 page 72 in text.) Make a histogram of survival times of 72 guinea pigs (Figure A.11) after they were injected with tubercle bacilli in a medical experiment.

Data Analysis

Analysis Tools

Anova: Two-Factor Without Replication
Correlation
Covariance
Descriptive Statistics
Exponential Smoothing
F-Test Two-Sample for Variances
Fourier Analysis
Histogram

OK
Cancel
Help

Figure A.12: Analysis ToolPak Dialog Box

**Solution.** Excel requires a contiguous block of data for the histogram tool.

1. Reenter the data in a block (cells A2:A73) and type the label "Times" in cell A1.

2. From the Menu Bar choose **Tools – Data Analysis** and scroll to the choice **Histogram** (Figure A.12). Click OK.

3. In the dialog box (Figure A.13) type the range A1:A73 in the **Input Range** area, which is the location on the workbook for the data. As with the Bar Chart you may instead click and drag from cell A1 to A73. The choice depends on whether your preference is for keyboard strokes or mouse clicks. Leave the **Bin Range** blank (because Excel will select the bins), check the **Labels** box because A1 has been included in the Input Range, type C1 for **Output Range** and check the box **Chart Output**. The option Pareto (sorted histogram) constructs a histogram with the vertical bars sorted from

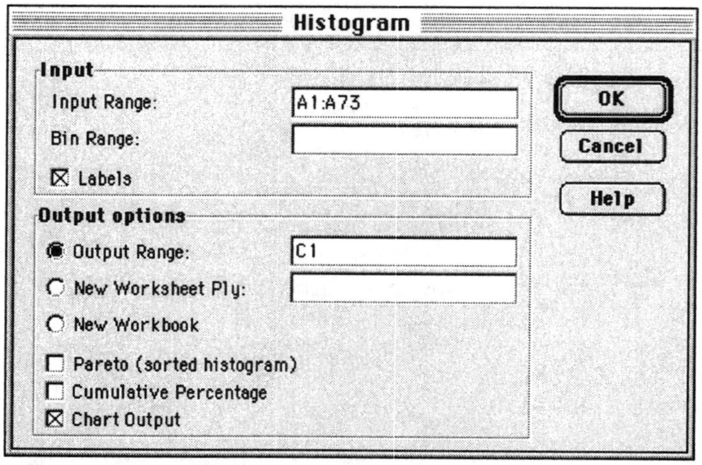

**Histogram**

Input
Input Range: A1:A73
Bin Range:
☒ Labels

Output options
◉ Output Range: C1
○ New Worksheet Ply:
○ New Workbook
☐ Pareto (sorted histogram)
☐ Cumulative Percentage
☒ Chart Output

OK
Cancel
Help

Figure A.13: Histogram Dialog Box

left to right in decreasing height. If Cumulative Percentage is checked, the output will include a column of cumulative percentages.

4. The output appears in Figure A.14. The bin interval boundaries (actually, the upper limit for each interval) appear in C2:C10 while the corresponding frequencies appear in cells D2:D10. The histogram appears to the right. We shall shortly modify the histogram by changing the labels and allowing adjacent bars to touch. But first, we explain how to customize the selection of bins.

| | C | D |
|---|---|---|
| 1 | *Bin* | *Frequency* |
| 2 | 43 | 1 |
| 3 | 123.625 | 45 |
| 4 | 204.25 | 16 |
| 5 | 284.875 | 4 |
| 6 | 365.5 | 1 |
| 7 | 446.125 | 2 |
| 8 | 526.75 | 2 |
| 9 | 607.375 | 0 |
| 10 | More | 1 |

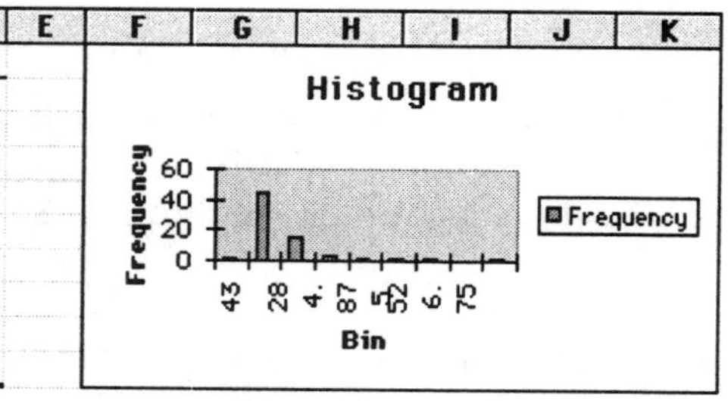

**Histogram**

Frequency
60
40
20
0
Bin
Frequency

Figure A.14: Output Table and Default Histogram

## Changing the Bin Intervals

If the bin intervals are not specified then Excel creates them automatically, so the number of bins roughly equals the square root of the number of observations beginning and ending at the minimum and maximum, respectively, of the data set.

In our example, we will first let Excel choose the default bins and then we will modify the histogram by selecting our own bin intervals.

1. Type "New Bin" (or another appropriate label) in cell B1. Then enter the values 100, 200, 300, 400, 500, 600, and 700 in B2:B8. An easy way to accomplish this is to type 100 and 200 in cells E2 and E3, respectively, then select E2 and E3, click the fill handle in the lower right corner of B3, then drag the fill handle down to cell B8, and release the mouse button.

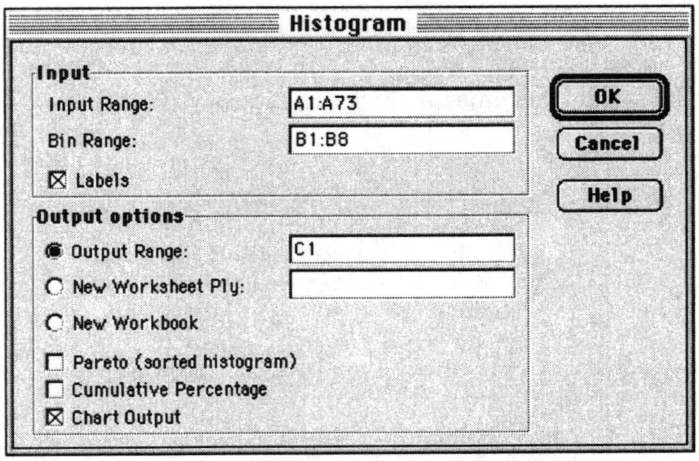

Figure A.15: New Bin Selection

2. Repeat the earlier procedure for creating a histogram, only this time type B1:B8 in the text area for **Bin Range** (Figure A.15). Before the output appears you will be prompted with a warning that you are overwriting existing data. Continue and the new output (Figure A.16) will replace Figure A.14 with your selected bin intervals.

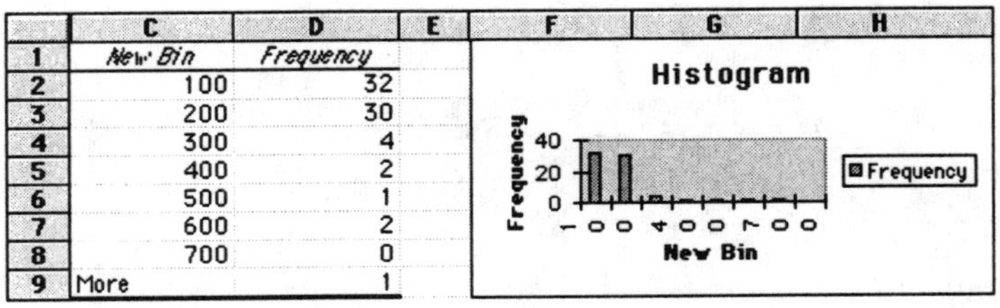

| | C | D |
|---|---|---|
| 1 | *New Bin* | *Frequency* |
| 2 | 100 | 32 |
| 3 | 200 | 30 |
| 4 | 300 | 4 |
| 5 | 400 | 2 |
| 6 | 500 | 1 |
| 7 | 600 | 2 |
| 8 | 700 | 0 |
| 9 | More | 1 |

Figure A.16: Output Table and New Bin Histogram

## Enhancing the Histogram

While the default histogram captures the overall features of the data set, it is inadequate for presentation. Excel provides a set of tools for enhancing the histogram. These are too numerous to all be mentioned here, but a few will be discussed with reference to the example. The other options may be invoked analogously.

**Resize.** Select the histogram by **clicking once** within its boundary and resize using the handles or move by dragging to a new location.

**Bar Width.** Adjacent bars do not touch in the default, which looks more like a bar chart for categorical data. To adjust the bar width, **double-click** the chart so the border becomes a thick grey cross-hatched line (Figure A.17).

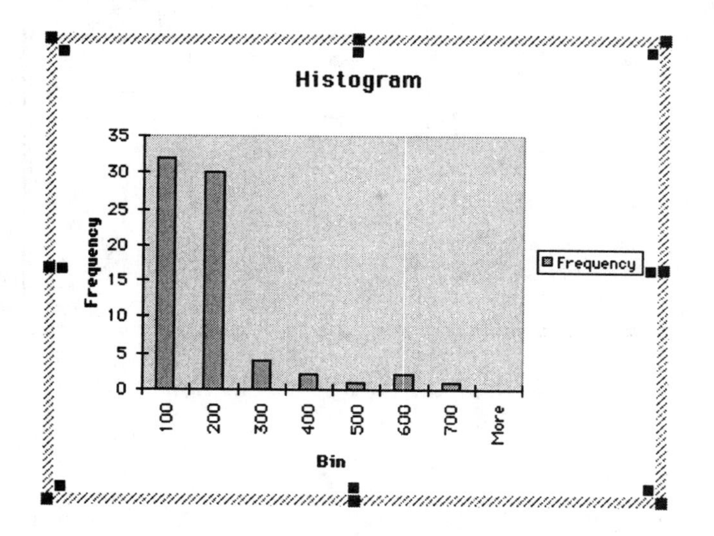


Figure A.17: Editing the Histogram

Select the X axis by clicking it once (Figure A.18). From the Menu Bar choose **Format – Column Group** and click the **Options** tab. Change the **GapWidth** to "0" and watch the histogram display change so adjacent bars touch (Figure A.19).

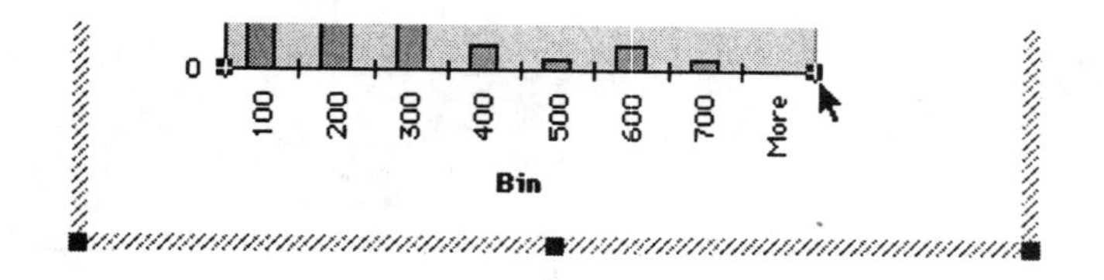


Figure A.18: Selecting the X Axis

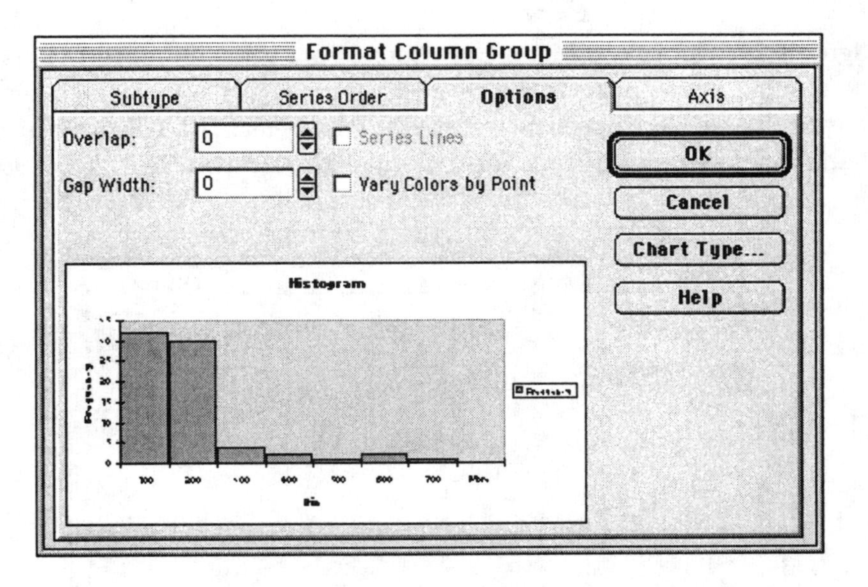


Figure A.19: Formatting Histogram Columns

**Chart Title.** Click on the title word "Histogram". A rectangular grey border with handles will surround the word, indicating it is selected for editing. Begin typing "Survival Times (Days) of Guinea Pigs," hold down the **Alt (Windows)** or the **Command (Macintosh)** keys and press enter. You may now type a second line of text in the **Formula Bar** entry area (Figure A.20). Continue typing "in a Medical Experiment," then click the Bold and the Italic buttons in the **Formatting Toolbar**.

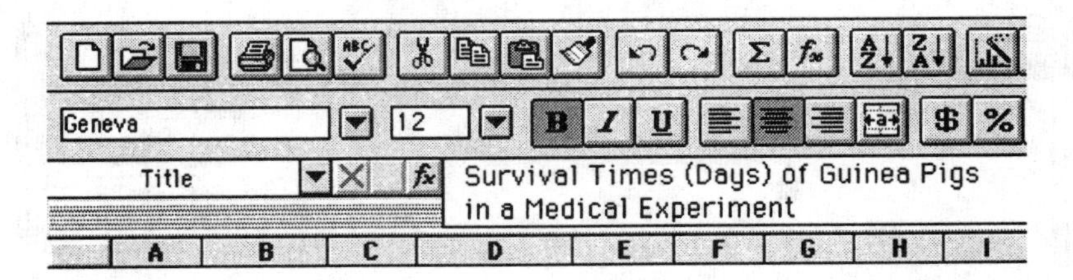


Figure A.20: Editing Chart Title

**X axis Title.** Click on the word Bin at the bottom of the chart (Figure A.17), type "Survival Time (Days)," and click Bold and Italic in **Formatting Toolbar**.

**Y axis Title.** Click on the word "Frequency" on the left side and then click Bold and Italic.

**X axis Format.** Select the X axis by clicking it once (as illustrated already in

FigureA.18). From the Menu Bar choose **Format – Selected Axis....** In the ensuing **Format Axis** dialog box (Figure A.21) you can click on various tabs to change the appearance of the X axis. Select the **Alignment** tab and change the orientation from Automatic to **horizontal** Text. Click OK.

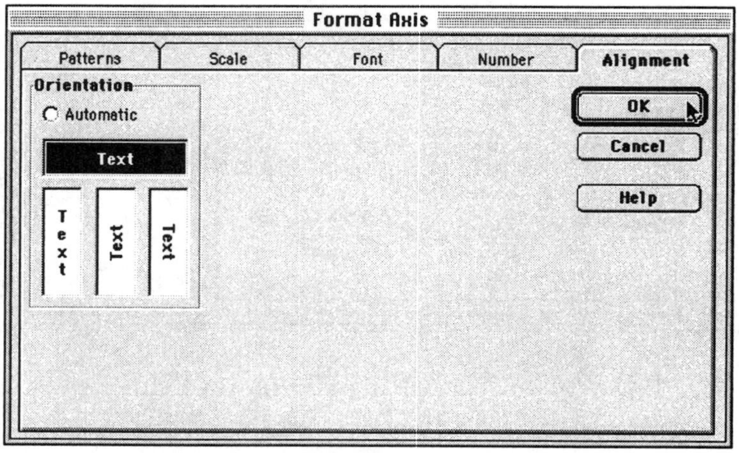

Figure A.21: Formatting the X Axis

**Y axis Format.** Select the Y axis by clicking it once. From the Menu Bar choose **Format – Selected Axis....** Click on the tab **Scale** and change the Maximum to 50 (Figure A.22). Click OK.

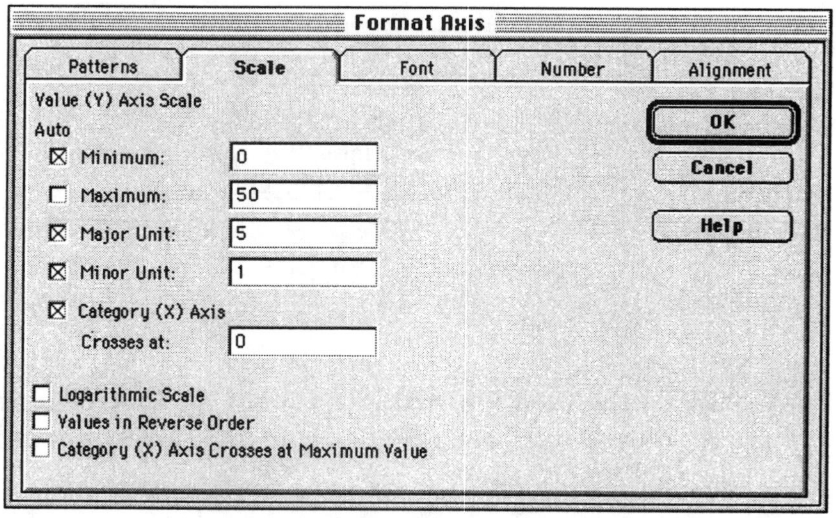

Figure A.22: Formatting the Y Axis

**Legend.** Remove the legend (which is not needed here) by clicking on it (the word "frequency" on the right in Figure A.17) and then pressing the delete key or choosing **Edit** – **Clear** – **All** from the Menu Bar.

**More Interval.** (Optional) Click any histogram bar, choose **Format** – **Selected Data Series...**, and in the **Format Data Series** box change the X values from $D$9 to $D$8, thereby removing the "More" interval. Click OK.

At the conclusion of the formatting, the histogram will appear as in Figure A.23.

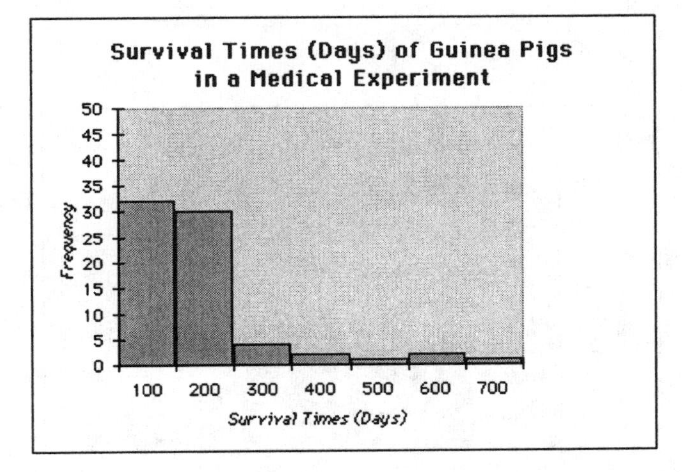


Figure A.23: Final Histogram after Editing

## Histogram from Grouped Data

The **Histogram** tool requires the raw data as input. When numerical data has already been grouped into a frequency table, it is the **ChartWizard** which is the appropriate tool. First use it to obtain a bar chart and then modify it exactly as you would enhance a histogram.

> **Example A.3.** (Figure 1.13 page 46 in text.) Figure A.24 gives the frequencies of vocabulary scores of all 947 seventh graders in Gary, Indiana, on the vocabulary part of the Iowa Test of Basic Skills. Column A is the bin interval and column B is the label for the histogram. To construct a histogram select B1:C12, click on the **ChartWizard**, then follow the steps presented earlier in Figures A.3–A.7. The final histogram is shown in Figure A.25.

| | A | B | C |
|---|---|---|---|
| 1 | Class | Bin | Number of Students |
| 2 | 2.0 - 2.9 | 3 | 9 |
| 3 | 3.0 - 3.9 | 4 | 28 |
| 4 | 4.0 - 4.9 | 5 | 59 |
| 5 | 5.0 - 5.9 | 6 | 165 |
| 6 | 6.0 - 6.9 | 7 | 244 |
| 7 | 7.0 - 7.9 | 8 | 206 |
| 8 | 8.0 - 8.9 | 9 | 146 |
| 9 | 9.0 - 9.9 | 10 | 60 |
| 10 | 10.0 - 10.9 | 11 | 24 |
| 11 | 11.0 - 11.9 | 12 | 5 |
| 12 | 12.0 - 12.9 | 13 | 1 |
| 13 | Total | | 947 |

Figure A.24: Vocabulary Scores

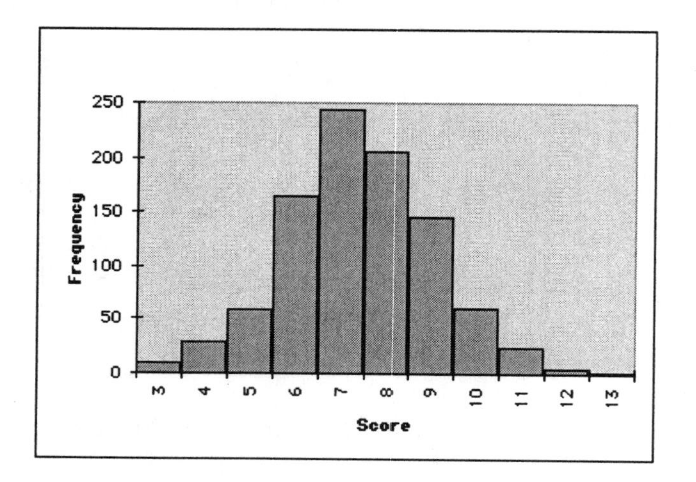


Figure A.25: Histogram from Grouped Data

## A.3 The Function Wizard

The **Function Wizard** assists in entering formulas and functions included in Excel particularly complex ones. The functions can perform decision-making, action-taking, or value-returning operations. The Function Wizard simplifies this process by guiding you step-by-step. All formulas are entered in the **Formula Bar** and the Function Wizard simplifies this process by guiding you step-by-step. There are three ways to invoke the Function Wizard: Click on the button in the **Standard Toolbar**; click on the button in the **Formula Bar** after you have begun to enter a formula; choose **Insert – Function** from the Menu Bar. With all three methods a dialog box **Function Wizard – Step 1 of 2** appears. Here you select a function from **Function Category** and **Function Name** lists. The definition and syntax of a function's arguments appear when a function is highlighted. There is a help button which describes the function in greater detail. After you select a function,

| | A | B | C | D |
|---|---|---|---|---|
| 1 | Beef Hot Dogs | | *Five Number Summary* | |
| 2 | | | | |
| 3 | Calories | | | |
| 4 | 186 | Min | 111 | =MIN(A4:A23) |
| 5 | 181 | Q1 | 140.5 | =QUARTILE(A4:A23,1) |
| 6 | 176 | Med | 152.5 | =MEDIAN(A4:A23) |
| 7 | 149 | Q3 | 177.25 | =QUARTILE(A4:A23,3) |
| 8 | 184 | Max | 190 | =MAX(A4:A23) |
| 9 | 190 | | | |
| 10 | 158 | | | |
| 11 | 139 | | | |
| 12 | 175 | | | |
| 13 | 148 | | | |
| 14 | 152 | | | |
| 15 | 111 | | | |
| 16 | 141 | | | |
| 17 | 153 | | | |
| 18 | 190 | | | |
| 19 | 157 | | | |
| 20 | 131 | | | |
| 21 | 149 | | | |
| 22 | 135 | | | |
| 23 | 132 | | | |

Figure A.26: Five-Number Summary

click Next for Step 2 of 2 in which you enter the arguments. These can be typed directly or *referenced* by using the mouse to point to data by clicking and dragging over cells.

We illustrate use of the **Function Wizard** by deriving the five-number summary of the calories data set shown in Figure 1.24 in Section 1.3.

## The Five-Number Summary

**Example A.4.** (See Exercise 1.39 page 42 in text.) Figure 1.24 shows the calories and sodium levels measured in three types of hot dogs: beef, meat (mainly pork and beef), and poultry. Find the five-number summary {minimum, first quartile, median, third quartile, maximum} for the calorie distribution of the hot dogs.

**Solution.** For illustration purposes we only consider the beef calories data.

1. Referring to Figure A.26 where we have entered the beef calorie data in cells A4:A23 of a workbook, enter the labels "Min," "Q1," "Med," "Q3," and "Max" in cells B4:B8. Select cell C5 to enter the quartile function.

2. Click on the **Function Wizard** button in the **Standard Toolbar** and select **Statistical** under Function Category and select `QUARTILE` under Function Name in the first dialog box. Click Next.

3. The second dialog box brings up the **QUARTILE** box. Enter A4:A23 for the data **array** and "1" for the **quart** to indicate the first quartile (Figure

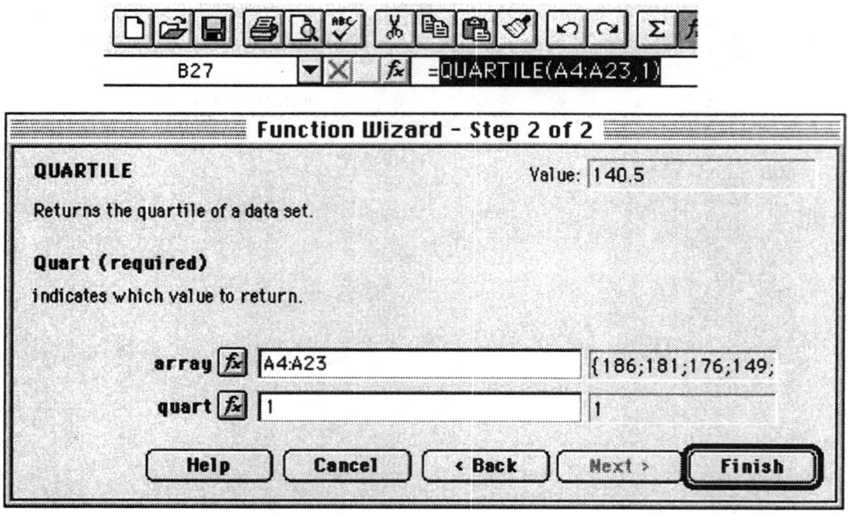

Figure A.27: Function Wizard Dialog Box – Quartile Function

A.27). The upper portion of Figure A.27 shows the completed formula, which appears automatically in the **Formula Toolbar**. Click Finish and the Function Wizard constructs the function and prints the value 140.5 in cell C5.

4. Continue with the rest of the five-number summary, either using the **Function Wizard** or entering the formulas by hand. Cells C4:C8 present the syntax while the values are in B4:B8.

The five-number summary is {111, 140.5, 152.5, 177.25, 190}. Note that Excel uses a slightly different definition of quartiles for a finite data set than the text.

## A.4 Scatterplot

Statistical studies are often carried out to learn whether or how much one measurement (an explanatory variable $x$) can be used to predict the value of another measurement (a response variable $y$). Once data is collected, either through a controlled experiment or an observational study, it is useful to examine graphically if any relationship is justified. We might plot in Cartesian co-ordinates all pairs $(x_i, y_i)$ of observed values. The resulting graph is called a **Scatterplot**.

**Example A.5** (The Endangered Manatee – Exercise 2.4 page 83 in text.) Manatees are large, gentle sea creatures that live along the Florida coast. Many manatees are killed or injured by powerboats. Table 2.1 contains data on powerboat registrations (in thousands) and

Table A.1: Manatees Killed and Powerboat Registrations

| Year | 1977 | 1978 | 1979 | 1980 | 1981 | 1982 | 1983 |
|---|---|---|---|---|---|---|---|
| Registrations (1000) | 447 | 460 | 481 | 498 | 513 | 512 | 526 |
| Manatees killed | 13 | 21 | 24 | 16 | 24 | 20 | 15 |

| Year | 1984 | 1985 | 1986 | 1987 | 1988 | 1989 | 1990 |
|---|---|---|---|---|---|---|---|
| Registrations (1000) | 559 | 585 | 614 | 645 | 675 | 711 | 719 |
| Manatees killed | 34 | 33 | 33 | 39 | 43 | 50 | 47 |

the number of manatees killed by boats in Florida in the years 1977 to 1990. Make a scatterplot of these data, labeling the axes with the variable names.

## Creating a Scatterplot

The steps involved in creating a scatterplot are similar to those for producing a **Histogram** using the **ChartWizard**, except that we use the **Scatterplot** chart type in the ChartWizard.

1. Enter the data from Table A.1 into cells B4:B17 and C4:C17 of a work book with the labels "Registrations (1000)" and "Manatees killed," referring to Figure A.31 later in this section.

2. Select cells B4:C17, click on the **ChartWizard**, and then click and drag the cross hair from one empty cell to another to select the rectangle in which the scatterplot will be displayed (block D1:I25 here).

3. Step 1. Dialog boxes appear (as in Figures A.3 – A.7 in this chapter) beginning with the ChartWizard dialog by asking you to confirm the data range selected. Click Next.

4. Step 2. Select **XY (Scatter)** as the chart type from the selection offered. Click Next.

5. Step 3. Specify Format **3** (Figure A.28).

6. Step 4. The dialog box (Figure A.29) shows a preview of the chart with the options selected. Confirm that the radio button for Data Series in **Columns** is selected, that the spin box for Column(s) for X Data contains "1," which tells Excel to use Column B for the $x$ (horizontally plotted) variable, and that "0" appears in the box Row(s) for Legend Text. A legend is not appropriate because each point represents two values $(x_i, y_i)$ (Figure A.29). Click Next.

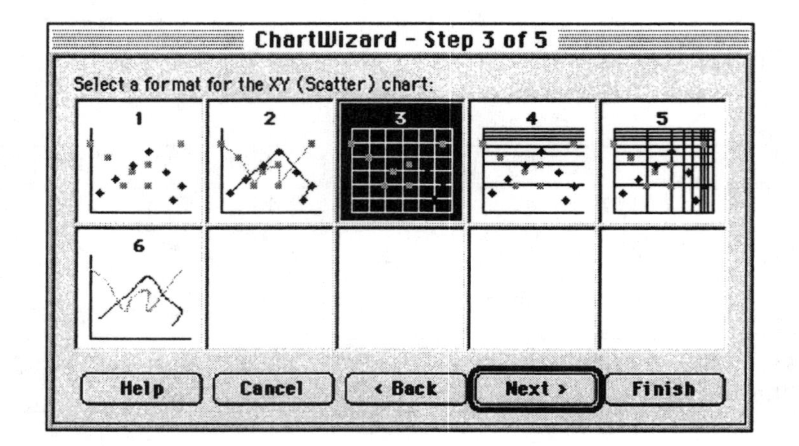


Figure A.28: Selecting Scatterplot in ChartWizard – Step 3

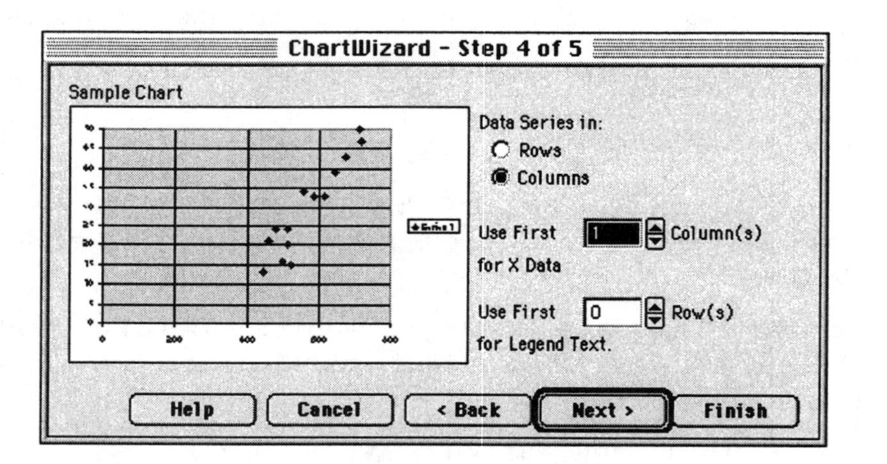


Figure A.29: Preview of Scatterplot – Step 4

7. Step 5. Select radio button **No** for Add a legend? and note that you can also change titles in this step.

8. The scatterplot appears embedded on your workbook (Figure A.31).

**Note:** Inspection of the scatterplots in Figures A.30 and A.31 reveals that the axes scales have been changed. Excel uses a range from 0 to 100% as the default and sometimes the scatterplot will show unwanted blank space, as is the case here. To remedy this, change the horizontal scale. Select the X axis by clicking it once, and from the Menu Bar choose **Format – Selected Axis...** and complete the dialog box. Similarly select the Y axis for editing. Refer to Figures A.21 and A.22 and the discussion for enhancing a histogram in Section A.2.

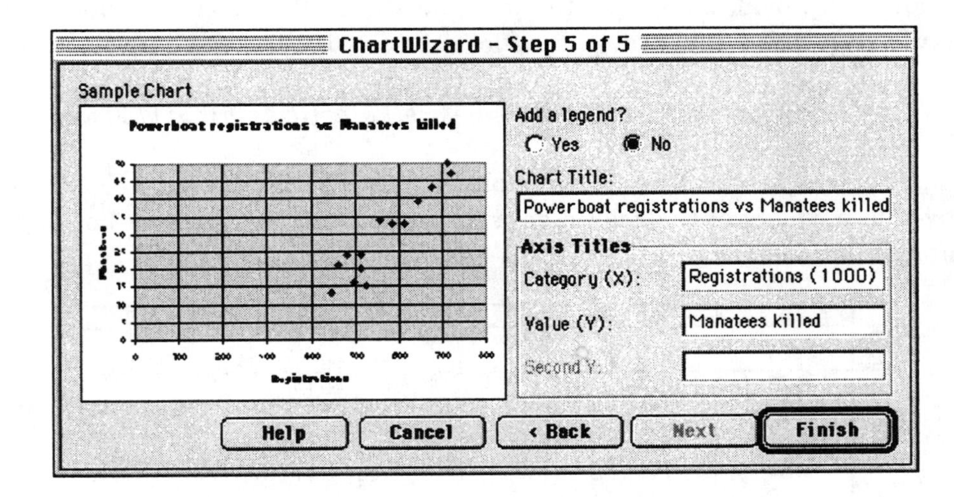


Figure A.30: ChartWizard – Step 5

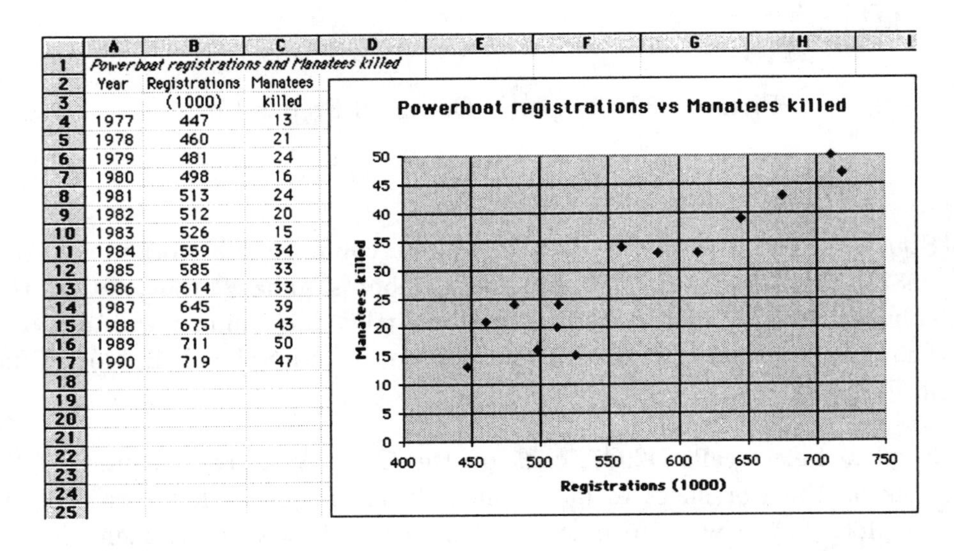

*Powerboat registrations and Manatees killed*

| Year | Registrations (1000) | Manatees killed |
|---|---|---|
| 1977 | 447 | 13 |
| 1978 | 460 | 21 |
| 1979 | 481 | 24 |
| 1980 | 498 | 16 |
| 1981 | 513 | 24 |
| 1982 | 512 | 20 |
| 1983 | 526 | 15 |
| 1984 | 559 | 34 |
| 1985 | 585 | 33 |
| 1986 | 614 | 33 |
| 1987 | 645 | 39 |
| 1988 | 675 | 43 |
| 1989 | 711 | 50 |
| 1990 | 719 | 47 |



Figure A.31: Manatee Data and Scatterplot Embedded on Sheet

## A.5 Boxplots

Excel does not provide a boxplot. However the Microsoft Personal Support Center has a web page "How to Create a BoxPlot – Box and Whisker Chart" located at

*http://support.microsoft.com/support/kb/articles/q155/1/30.asp*

with instructions for creating a reasonable boxplot. We illustrate by constructing side-by-side boxplots of the calorie data for beef, meat, and poultry hot dogs in Example A.3.

| | A | B | C | D | E | F | G |
|---|---|---|---|---|---|---|---|
| 1 | | | *Boxplots of Hot Dog Calories* | | | | |
| 2 | Beef | Meat | Poultry | | Beef | Meat | Poultry |
| 3 | 186 | 173 | 129 | median | 152.5 | 153 | 129 |
| 4 | 181 | 191 | 132 | Q1 | 140.5 | 139 | 102 |
| 5 | 176 | 182 | 102 | min | 111 | 107 | 86 |
| 6 | 149 | 190 | 106 | max | 190 | 195 | 170 |
| 7 | 184 | 172 | 94 | Q3 | 177.25 | 179 | 143 |
| 8 | 190 | 147 | 102 | | | | |
| 9 | 158 | 146 | 87 | | | | |
| 10 | 139 | 139 | 99 | *Formulas for Beef Column* | | | |
| 11 | 175 | 175 | 170 | median | =MEDIAN(A3:A22) | | |
| 12 | 148 | 136 | 113 | Q1 | =QUARTILE(A3:A22,1) | | |
| 13 | 152 | 179 | 135 | min | =MIN(A3:A22) | | |
| 14 | 111 | 153 | 142 | max | =MAX(A3:A22) | | |
| 15 | 141 | 107 | 86 | Q3 | =QUARTILE(A3:A22,3) | | |
| 16 | 153 | 195 | 143 | | | | |
| 17 | 190 | 135 | 152 | | | | |
| 18 | 157 | 140 | 146 | | | | |
| 19 | 131 | 138 | 144 | | | | |
| 20 | 149 | | | | | | |
| 21 | 135 | | | | | | |
| 22 | 132 | | | | | | |

Figure A.32: Boxplot – Data and Preparation

1. Step 1. Enter the calorie data into three columns of a workbook (Figure A.32), then find and enter the five-number summary into another three columns **in the order** median, first quartile, minimum, maximum, third quartile. We have entered this information, including labels in block D3:G7 in Figure A.32).

2. Step 2. Select cells D2:G7, click on the **ChartWizard** button, and then click on the worksheet to locate the cell for the upper left corner of your boxplot. If you want to place the boxplot on a new sheet, then after you have selected D2:G7, choose **Insert – Chart – As New Sheet** from the Menu Bar. We have located the boxplot to begin in cell H1. Click Next (Figure A.33).

3. Step 3. Select the **Combination** chart type and click Next.

4. Step 4. Select chart style number **6** and click Next. An Alert Box appears with the following warning: **A volume-open-high-low-close stock chart must contain five series.** Click OK.

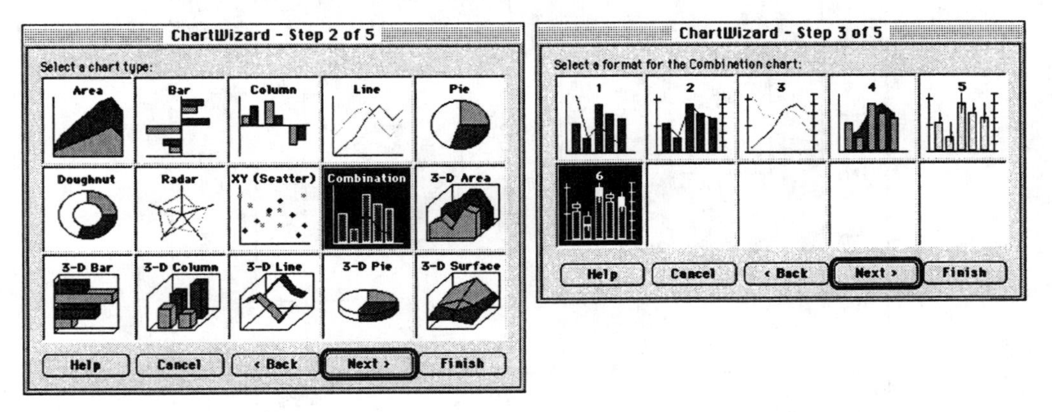


Figure A.33: Boxplot – ChartWizard Steps 2 and 3

5. Step 5. In the next dialog box check the radio button **Rows** for Data Series in: and click Next (Figure A.34).

Figure A.34: Boxplot – ChartWizard Steps 4 and 5

6. Step 6. In the final dialog box check the radio button **No** for Add a legend? and add a Chart Title "Boxplots of Calories Data."

Next we edit this chart.

1. Double-click the chart to activate it. From the Menu Bar choose **Insert – Axes** and clear the check box next to **Value (Y) Axis** under **Secondary Axis**. Click OK.

2. Click once on any one of the colored columns to select the series. Do not click one of the white columns. From the Menu Bar choose **Format – Chart**

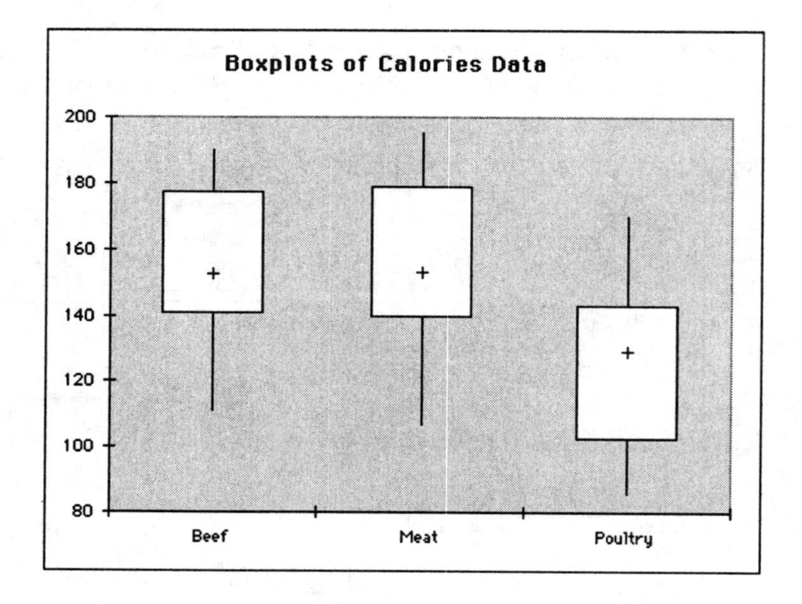


Figure A.35: Boxplot of Calories Data

**Type...**, click **Line**, and then click OK. A line that connects the three white columns appears in the chart.

3. Click once on the line and from the Menu Bar choose **Format – Selected Data Series....**

4. Under the **Patterns tab**, select **None** for **Line** and **Custom** for **Marker.** For the custom marker choose the plus sign from the **Style** list, the color black from the **Foreground** list, and **None** from the **Background** list. Click OK.

5. Double-click the **Y axis** and under the **Scale** tab set the Minimum to 80 and click OK. The final boxplot appears on your sheet (Figure A.35).